Archibald Liversidge

Mines and Mineral Statistics of New South Wales,

And Notes on the Geological Collection of the Department of Mines

Archibald Liversidge

Mines and Mineral Statistics of New South Wales,
And Notes on the Geological Collection of the Department of Mines

ISBN/EAN: 9783744677257

Printed in Europe, USA, Canada, Australia, Japan

Cover: Foto ©berggeist007 / pixelio.de

More available books at **www.hansebooks.com**

New South Wales Intercolonial and Philadelphia International Exhibition.

MINES AND MINERAL STATISTICS

OF

NEW SOUTH WALES,

AND

NOTES ON THE GEOLOGICAL COLLECTION OF THE DEPARTMENT OF MINES,

COMPILED BY DIRECTION OF

THE HON. JOHN LUCAS, M.P.

MINISTER FOR MINES.

ALSO

EMARKS ON THE SEDIMENTARY FORMATIONS OF N. S. WALES,

BY

THE REV. W. B. CLARKE, M.A., F.G.S., F.R.G.S.,

MEMBER OF THE GEOLOGICAL SOCIETIES OF FRANCE AND AUSTRIA,

AND

OTES ON THE IRON AND COAL DEPOSITS, WALLERAWANG, AND ON THE DIAMOND FIELDS,

BY

PROFESSOR LIVERSIDGE, F.C.S., F.G.S.,

SYDNEY UNIVERSITY.

SYDNEY: THOMAS RICHARDS, GOVERNMENT PRINTER.

1875.

TABLE OF CONTENTS.

MINERAL STATISTICS.

As the result of an attempt to place before the public a statement of the quantity and value of the minerals won from the

CORRIGENDA.

Page	Paragraph	For	Read
72	third	"*Coinet*"	Comet
85	sixth	"*dense*"	druse
156	sixth	"*Palæzoic*"	Palæozoic
160	last	"*on*" N.S.W.	in
174	second	"*over*"	occur
177	second	"*Palæzoic*"	Palæozoic
193	fourth	"*prints*"	plants

total value of the several minerals raised to the end of 1874, and also in the preparation of the other statements submitted.

	£	s.	d.
Gold, total value...	30,536,246	10	6
Coal, do.	6,655,328	0	0
Tin, do.	866,461	0	0
Copper, do.	807,476	0	0
Shale, do.	261,414	0	0
Silver, do.	77,216	0	0
Iron, do.	15,434	0	0
Antimony, do.	897	0	0
	£39,220,472	10	6

MINERAL STATISTICS.

As the result of an attempt to place before the public a statement of the quantity and value of the minerals won from the mines of this Colony, the subjoined tables are submitted. That these tables are so incomplete and represent only approximately the produce of our mines is due to the facts that, owing to the recent establishment of this department, the arrangements for collecting statistical information have not yet been fully completed, that previous to its establishment the means of collecting and preserving such information were most imperfect, and that after the lapse of a great number of years it is most difficult to gather information which could at the time have been obtained with comparative ease. The value of complete statistics brought up to date can scarcely be overrated, because they would form the basis of future labours, and for this reason past neglect of this important subject is much to be regretted. The arrangements for the future are such as it is hoped will secure the collection, publication, and preservation of complete and authentic returns, and no pains will be spared to render our records of the past more complete than they are at present.

The statistics published annually by the Registrar General, and the information kindly supplied by the Collector of Customs and the owners and managers of collieries, and of some of our copper and tin mines, have been of the greatest assistance in the preparation of the following statement, which shows the total value of the several minerals raised to the end of 1874, and also in the preparation of the other statements submitted.

	£	s.	d.
Gold, total value...	30,536,246	10	6
Coal, do.	6,655,328	0	0
Tin, do.	866,461	0	0
Copper, do.	807,176	0	0
Shale, do.	261,414	0	0
Silver, do.	77,216	0	0
Iron, do.	15,434	0	0
Antimony, do.	897	0	0
	£39,220,472	10	6

GOLD.

Considering the vast extent of the auriferous deposits of this Colony, it will at first sight appear strange that the quantity of gold raised since the first discovery (now some twenty-four years ago) should amount to only 8,205,232 ounces. It must be stated that the total quantity given represents only the gold passed through the Custom House and received at the Mint, and does not include the gold used in this Colony in manufactures and otherwise, and sent out of the Colony by private hands before the imposition of the import duty. In view of the fact that, at a time when gold was purchased simply by weight, a higher price could be obtained in an adjoining Colony than in this, there does not appear to be any reason for supposing that any gold raised in that Colony is included in the above total, but there are cogent reasons why that total may not include all the gold got in this Colony.

In the absence of statistics showing the average yields obtained from given quantities of the alluvial and vein stuff of our several Gold Fields, the total given (supposing it to represent even approximately the quantity of gold raised in this Colony) should by no means be regarded as evidence that our gold deposits are not both rich and extensive. Those acquainted with the Gold Fields of this Colony will accept the smallness of the quantity of gold raised as indicating that our auriferous resources have not only not been fully developed but have scarcely been touched. Except in some few localities the quartz veins have not been worked to any depth, and may be said to be practically untested. Alluvial lands have in some instances been worked to a depth of some 200 feet, but there are the strongest indications of the existence of deep leads in various parts of the Colony where no attempt has yet been made to work them; and comparing the work that has been done in this branch of mining with what has to be done, the working of deep leads may be said to have scarcely commenced. Indeed, up to the present time (some few localities excepted) mining has been limited to the working of river beds and shallow alluvial deposits, and even this has been done in the roughest manner and with the aid of the rudest appliances. That greater strides have not been made in the discovery of new deposits, and in the development of those already known, may be due in a very great measure to the sparseness of our mining population. In the case of Parkes and Gulgong, where large numbers of miners are located, new discoveries are frequent, and prospecting is carried on with some degree of system, but on the thinly populated fields the miners appear to be able to make a living by working or reworking old ground in the old fashion, and so long as they can make a moderate income they seem to have no incentive to search for new deposits, or for appliances by means of which the ground might be worked more effectively. It is matter for congratulation that attempts are being made here and there to introduce a system of mining on a large scale, and with such appliances as will render the working of poor ground highly remunerative; and as the results of such enterprises come to be known and appreciated, persons will be found ready to embark in like ventures, and thus our gold yields will be largely increased.

It is hoped that as soon as the arrangements of the Mining Department shall be complete it will be enabled to make known the vast extent of the auriferous deposits of this Colony, and that a large number of experienced miners will be thereby attracted to our Gold Fields. Every effort will be made in future to ascertain and publish returns showing the average yield of gold from the alluvions and vein stuffs of our several Gold Fields, because such returns may be regarded as the best test of the value of our auriferous deposits, and—allowance being made for the peculiar conditions attending the working of the several kinds of deposits—a valuable guide to miners and investors. That there is ample scope for the profitable employment of an almost unlimited number of miners in opening up auriferous country yet unprospected, and in working deposits now neglected for want of the requisite appliances and skill, cannot be doubted.

In many instances the small size of our quartz veins and the hardness of the country render quartz-mining comparatively unremunerative, but in a great number of instances the abandonment or neglect of our quartz veins is due to want of experience and want of proper appliances rather than to the poorness of the stone. Again, the combination or association of other metals with the gold tends to lessen the value of the gold, and in the absence of the means of separating and saving the whole of these metals, and the skill to employ such means to the greatest advantage, some veins are regarded as unremunerative which, under skilful management, would prove highly remunerative.

Up to the present time comparatively little attention has been paid to the saving and treatment of pyrites. In many instances quartz heavily charged with pyrites is crushed at a battery where there are no appliances for saving them, and where in fact no attempt is made to save them. The pyrites are frequently carried off with the tailings into a creek and lost. The loss sustained by the waste of such material, though there are no means of estimating its extent, must be very great, and what is worse, the loss may be said to be irreparable, because it would be impossible to collect the material again. If the gold thus lost could have been saved, many claims abandoned as unremunerative might have been worked with profit for years to come.

In a very large number of instances where the miners do not crush their own stone, the cost of carting and crushing amounts to about twenty-five shillings, equal to between six and seven dwts. of gold per ton, so that unless the stone is moderately rich it cannot pay for working. The miner has, however, not only to submit to the loss of the gold contained in the pyrites, and to pay, say six or seven dwts. of gold for having his stone crushed, but in many cases the battery is in the charge of a person who has no special knowledge of and experience in the reduction of quartz and the saving of gold, and all the various descriptions and qualities of stone are subjected to precisely the same treatment. It would not, perhaps, be practicable for every company, nor for the owner of every crushing plant, to erect the apparatus requisite for the collection and effective treatment of pyrites, but it is highly desirable in the interest of mining that the pyrites should be saved, because if it were known that the miners were in the habit of saving them, a market would soon spring up in which they would be able to obtain nearly as much money from the sale of their pyrites as they would gain by simply extracting the gold from them, and the practice of saving these waste products would lead to the erection of works for the extraction and saving not only of the gold, but also of the sulphur, arsenic, &c.

The following brief extracts from the Report of the Board recently appointed in Victoria to report upon the methods of treating pyrites and pyritous vein-stuffs are submitted for the information of those who may not have access to the report:—"It is decidedly better to crush quartz containing pyrites raw. The great majority of witnesses are in favour of using Borlase's Concave Buddle with Munday's Patent Scrapers for the purpose of separating the various descriptions of pyrites from the crushed material. An 18 to 24 feet machine making seven or eight revolutions per minute is recommended. That it is absolutely necessary to roast pyrites previous to amalgamation, and for this purpose reverberating furnaces, with inclined hearths, are the best at present in use in Victoria. That the introduction of combustible substances, such as charcoal, into the furnace with the pyrites is not advisable, and that attention should be given to the regular supply of fuel and to the proper regulation of the draught. That grinding the roasted pyrites in Chilian Mills, Arrastras, or Wheeler's Pans, is considered the best mode of amalgamating. The witnesses are said to be unanimous as to the absolute necessity of thoroughly breaking up the quicksilver, in order that it may penetrate the stuff operated on and gather up all the gold brought in contact with it, but the very means which are necessary to secure effective amalgamation give rise to a form of floured mercury brought about by mechanical (not chemical) action. The evil is said to be as great in this as in the case where chemical combinations retard the saving of gold, for it appears that during the operation of flushing off part of the floured mercury (equal, say, to 2½ lbs. of quicksilver per ton of roasted ore treated), is carried way with the water and lost. This, in addition to a considerable loss of gold as amalgam, renders it evident that if some effectual means of saving both be brought into general use many thousands of tons of tailings now lying unworked, which it would not otherwise pay to operate upon, could be made to yield a considerable profit and open up a large field of labour."

Under the head of waste products the Board says :—" Very little attention has been given in Victoria to the extracting other constituents of pyrites, such as silver, copper, nickel, cobalt, and other metals, as also sulphur and arsenic, which from a purely commercial point of view, are of vast importance, and capable of adding materially to the wealth of the Colony. Most of the waste products, as they are termed, are such no longer, as their presence has increaed the value of and demand in Europe for pyrites of all kinds." Messrs. Johnson, Matthey, & Co., assayers, of London, writing on this subject, say, referring to some samples of auriferous pyrites submitted to them by the Port Phillip and Colonial Gold Mining Company :—' We beg to remark that mineral of this character would find a ready and advantageous market in England, provided that a regular quantity would be shipped.' Our experience certainly leads us to the conclusion that as the working of these ores become better understood in England (a result immediately consequent upon a steady supply), a keen competition will arise and an extreme value be realized."

The Board says :—" We are fully of opinion that the establishment of large central works for the treatment of pyrites on the most thorough and approved system would be a very great advantage to the Colony (Victoria) in every respect. Such works would give rise to the foundation of many new industries, besides tending to the production of large quantities of gold. It would be necessary to establish works in situations easily accessible to railway communication from the Gold Fields, where the supply of raw pyrites would be regularly and easily obtained. There could be no objection to the occupation by such establishments of advantageous sites on the ground of injury to public health, as the fumes given off in the process of roasting pyrites would be condensed for the purpose of saving the sulphur and arsenic." It is said that more than 72,000 tons of pyrites and pyritous vein stuffs are treated in Victoria every year. And the Board says "tens of thousands of pounds are annually lost to that Colony by the non-existence of such establishments as are advocated."

Mr. Henry Boehme, of Ashfield, who has for some time past been making experiments with a view to discover some more economical and effective method of reducing quartz, is of opinion that he has succeeded. His process consists in burning the quartz in a closed furnace, and when a certain heat is reached the quartz is taken from the furnace and thrown into water. The following extracts from the Report of the Victorian Pyrites Board on the subject of burning quartz before crushing may be of interest. "It is asserted that it would be impossible to thoroughly oxidize the sulphur by burning, and that a lower sulphide would result which would melt and enclose the fine particles of gold, rendering the subsequent extraction more difficult. Mr. Latta's evidence, which is of a practical nature, is very decisive in regard to the question. He states that roasting has the effect of converting the gold in the exterior of the quartz into globules, and covering them with a ferruginous glaze, which is prejudicial to the proper extraction of the precious metal. He has microscopically examined raw and burnt quartz washings from the lowest blankets and found most particles of gold in the latter. Mr. Heywood considers that many substances are removed during the process of roasting which would tend to interfere with amalgamation, and that the loss of mercury is less. Mr. Kayser thinks that fine gold is run into a globular form during roasting, and is then more easily amalgamated. On the other hand, Mr. John Lewis, the Manager of the New North Clunes Company, is very emphatic in his condemnation of burning, for he says,—'I would burn quartz previous to crushing if the pyrites were not to be saved, but if they were needed I would crush raw.' After a careful consideration of this portion of the question, the Board is of opinion that crushing raw is the best method of treatment except when pyrites are absent, then burning might be adopted if fuel is easily obtainable in order to economise wear and tear of machinery in crushing." Mr. Boehme however claims that his process is not open to any of the objections urged against the roasting of quartz. He says, the pyrites contained in the quartz are not prejudicially affected by the process, and that the whole merit of the process consists in having discovered the exact heat to which quartz may be subjected without producing the effects referred to. He has forwarded samples of raw and roasted quartz to this department for exhibition. The recent arrival of these samples has rendered an examination for the purpose of testing the merits of his process impossible, but he offers the fullest information to those interested, and courts investigation, so that there is every reason to hope that the merit of the process, if it possess any, will speedily be made public.

Before leaving the subject of quartz-mining, reference may be made to the evil effects upon this branch of mining of the want of system in the mode of working vein stuffs, and the consequence of neglecting to keep prospecting works well ahead.

It frequently happens that, where a vein of rich stone is cut, miners inexperienced in working quartz lodes take out all the gold-bearing stone in view, and then abandon the vein as worthless; whereas, if they had opened out their mine systematically, and kept their prospecting works well in advance, they might have cut another or other equally rich seams, and then would have secured a permanently profitable mine. Many veins so opened and abandoned are left unworked for a very long time, being passed by as worthless by miners in search of a claim. Sometimes, as it were, by the merest chance, a party of miners set to work in such a claim, and reap the reward abandoned by their predecessors. Many veins in which the gold in very fine particles is evenly distributed through the stone are regarded as worthless because the gold is not visible: such veins will yet be profitably worked as our experience increases and our facilities become greater.

ESCORT RETURNS

Name of Gold Field.	1858.	1859.	1860.	1861.	1862.	1863.	1864.	1865.
Western District.								
Sofala	43619·48	35737·97	42727·97	38307·10	35793·62	30167	26266·30	31855·7
Bathurst	10035·08	11714·51	23566·34	28614·52	23872·72	28680	17679·30	21989·9
Hargreaves	40685·50	37475·91	31009·72	21635·02	7888·06	11061	8167·80	5892·11
Tambaroora	24157·38	16367·45	18861·67	22997·94	18956·35	21836	16355·99	15246·6
Mudgee	14694·57	21814·40	10597·32	10281·50	16626·65	21194	19848·87	19782·4
Orange	3734·99	2891·89	2813·22	3943·14	783·36	11744	25022·43	22751·7
Stoney Creek	11164·88	4765·08	5757·92	4444·30	5251·84	19268	17646·74	13863·6
Forbes		......	......	4451·10	212547·84	71493	18722·31	9869·2
Weddin		......	......	......	......	......	......	......
Grenfell		......	......	......	......	......	......	......
Carcoar		......	......	......	......	......	......	......
Southern District.								
Goulburn	5263·49	8868·47	7472·81	2371·09	1577·24	4083	4056·23	2867·04
Braidwood	37278·89	56179·75	65755·10	68016·25	60226·54	63736	60521·82	67846·3
Tumut	29622·06	41927·98	4408·48	......	2661·60	6066	1168·60	1293·0
Gundagai	23·70	251·54	140·82	701·87	548·89	364	6260·27	6732·6
Meragle	100·45	15·57	......	......	......	......	......	......
Tumberumba		1133·31	959·42	684·35	......	......	1533·65	2868·9
Adelong		......	28082·37	24751·62	20125·17	16498	18720·25	13003·4
Kiandra		......	67687·03	16565·76	7385·49	6870	6866·94	4231·4
Cooma		......	1424·48	512·57	718·71	1021	147·82	282·5
Yass		......	4032·34	9510·47	20·92	13	116·78	......
Burrangong		......	......	109879·05	124648·69	60142	32411·14	17592·6
Nerrigundah		......	......	......	7662·81	11973	7454·54	600·5
Araluen		......	......	......	......	......	......	......
Northern District.								
Rocky River	17277·06	16102·21	9053·51	12091·75	7514·90	13748	9946·66	5704·49
Nundle	11022·04	20514·36	11307·86	11890·08	9192·92	9508	8476·71	8123·2
Tamworth	3474·17	2900·32	2490·22	2609·67	2133·20	4257	2399·37	2990·2
Timbarra		18428·48	45423·80	8482·27	8130·79	8953	6195·23	4178·1
Grafton		......	......	102·71	26·88	56	......	......
Scone		......	......	......	1243·08	706	437·84	921·3
Armidale		......	......	......	......	......	......	322·5
Total	252382·74	387768·30	356472·38	402634·13	575538·37	423407	316429·59	280810·1

OF GOLD.

1866.	1867.	1868.	1869.	1870.	1871.	1872.	1873.	1874.	Total Ozs.	
22876·67	20959·85	16954·88	15779·34	12903·16	12134·97	10765·15	9507·90	9372·65	415729·78	
26899·39	18594·10	14018·60	15816·05	27579·38	8574·99	9805·82	9738·39	10476·90	307856·04	
6404·79	5654·80	4903·32	4387·53	3753·15	4512·43	4100·40	4448·35	3603·40	205585·29	
17305·52	14390·26	14000·06	17567·10	18698·83	42073·21	80592·46	62834·48	25066·18	447707·52	
18601·25	16366·86	24090·37	20177·33	19966·63	87653·79	140538·81	131124·77	75684·74	659044·28	
12546·49	3142·72	4975·66	10431·29	6078·13	9256·55	7679·01	5895·98	7170·94	140870·56	3001062·90.
9457·35	5511·74	4679·08	3669·19	3424·38	3104·92	4405·79	5285·11	2457·37	124157·19	
16503·07	7746·44	3067·19	1583·41	……	……	……	628·92	532·67	346845·22	
240·26	……	……	……	……	……	……	……	……	240·26	
……	42081·55	45254·82	41542·69	36230·95	21979·00	36413·55	32729·82	50011·97	306244·35	
……	……	……	……	……	20104·06	12964·04	6224·34	7489·97	46782·41	
2540·17	1621·82	1075·36	539·60	350·12	2592·19	1827·10	364·28	308·10	47778·08	
46165·77	20422·73	7837·62	11170·42	12411·56	20665·37	15644·51	10086·88	10320·85	635086·40	
1291·31	503·76	453·94	173·69	……	2522·80	4282·26	2854·81	2331·56	101562·27	
7918·67	6884·37	1889·62	2847·67	3814·66	2696·83	874·30	1809·50	2331·61	46093·00	
……	……	……	……	……	……	……	……	……	136·02	
2354·21	1754·12	1473·29	1310·36	1212·78	4696·24	3719·87	825·63	217·32	24743·53	
11247·62	10731·53	12901·19	12367·73	10772·08	13825·70	17042·41	21607·78	17850·60	249527·50	1866633·67.
3487·04	2669·92	2234·67	2419·65	1651·33	1811·87	648·10	……	……	124529·24	
1472·64	2293·49	693·79	2536·11	1261·48	528·32	2293·19	2091·89	1421·82	18699·85	
……	42·67	……	……	……	……	……	……	48·01	13784·19	
12332·91	15249·41	8325·11	8399·74	6974·86	6285·29	8175·48	3815·42	3496·49	418028·19	
……	……	……	……	……	……	……	……	……	27700·92	
……	6767·24	46634·23	38824·59	17309·08	17637·76	20000·59	7236·58	4554·41	158964·48	
6093·70	4274·12	3144·77	2734·43	3121·81	2040·92	1569·86	2242·18	2155·83	118813·34	
6936·93	4762·82	3774·60	5905·41	6347·40	7238·97	4399·38	3862·43	3357·86	139821·00	
4079·46	5100·46	3652·88	950·09	1308·74	1395·53	1497·84	1169·82	1057·72	43556·73	
1567·12	……	……	……	……	……	……	……	……	71058·84	405685·69.
……	……	……	……	……	……	……	……	……	155·59	
669·66	332·69	468·47	521·11	……	……	68·59	45·70	……	5434·51	
2497·47	4835·81	3235·39	2727·74	3495·84	3587·24	2575·94	1766·07	1798·71	26840·68	
241489·47	222715·29	229738·82	224382·27	198664·32	296928·01	392186·06	323197·03	243518·13	5273382·26	

GOLD.

Year.	Received at Mint.	Exported without passing through Mint.	Total Quantity.	Value of Gold received at Mint.	Value of Gold Exported.	Total Value.
	ozs.	ozs.	ozs.	£ s. d.	£ s. d.	£ s. d.
1851	....	144,120·808	144,120·808	....	468,336 0 0	468,336 0 0
1852	....	818,751·93	818,751·93	....	2,660,946 0 0	2,660,946 0 0
1853	....	548,052·99	548,052·99	....	1,781,172 0 0	1,781,172 0 0
1854	....	237,910·69	237,910·69	....	773,209 0 0	773,209 0 0
1855	106,983	64,384	171,367	408,656 0 0	245,938 0 0	654,594 0 0
1856	142,137	42,463	184,600	530,598 0 0	158,576 0 0	689,174 0 0
1857	126,950	48,999	175,949	487,439 0 0	187,038 0 0	674,477 0 0
1858	272,825·65	13,973	286,798·65	1,050,339 12 2	53,835 0 0	1,104,174 12 2
1859	308,183·41	21,180	329,363·41	1,178,114 7 10	81,013 0	1,259,127 7 10
1860	352,222·780	31,831·13	384,053·910	1,341,430 10 7	123,942 9 2	1,465,372 19 9
1861	414,861·84	50,823·53	465,685·37	1,608,277 8 4	197,894 2 4	1,806,171 10 8
1862	587,895·24	52,727·55	640,622·79	2,262,471 18 2	205,307 17 11	2,467,779 16 1
1863	401,713·25	64,398·38	466,111·63	1,545,419 0 2	250,751 3 10	1,796,170 4 0
1864	308,670·64	31,596·38	340,267·02	1,181,897 19 10	123,028 8 1	1,304,926 7 11
1865	300,277·57	20,039·16	320,316·73	1,153,215 8 0	78,027 9 7	1,231,242 17 7
1866	269,239·71	20,774·5	290,014·21	1,035,513 0 3	80,890 14 2	1,116,403 14 5
1867	259,556·92	12,329·55	271,886·47	1,005,569 19 3	48,008 3 8	1,053,578 2 11
1868	232,488·660	23,173·72	255,662·380	904,422 7 0	90,232 13 5	994,655 0 5
1869	179,549·24	71,942·55	251,491·79	694,022 7 4	280,125 6 0	974,148 13 4
1870	143,709·44	97,148·66	240,858·10	552,743 16 8	378,272 11 10	931,016 8 6
1871	242,596·91	81,012·88	323,609·79	935,040 17 11	315,443 18 0	1,250,484 15 11
1872	331,273·19	93,856·72	425,129·91	1,278,127 4 11	365,454 12 0	1,643,581 16 11
1873	259,579·77	102,204·94	361,784·71	997,214 18 11	397,960 9 8	1,395,175 8 7
1874	233,216·59	37,606·72	270,823·31	895,147 10 3	145,181 3 3	1,040,328 13 6
Total	5,473,930·810	2,731,301·788	8,205,232·598	21,045,661 7 7	9,490,585 2 11	30,536,246 10 6

VALUE per Ounce of Gold from the several Gold Fields.

Stations.	1858.	1859.	1860.	1861.	1862.	1863.	1864.	1865.	1866.	1867.	1868.	1869.	1870.	1871.	1872.	1873.	1874.	Average.
	s. d.	s. d.	s. d.	s. d.	s. d.	s. d.	s. d.	s. d.	s. d.	s. d.	s. d.	s. d.	s. d.	s. d.	s. d.	s. d.	s. d.	£ s. d.
W.—Sofala	77 1	77 1	77 3	76 11	77 2	76 4	76 11	76 8	77 0	76 1	75 8	75 8	76 5	76 8	77 2	77 1	76 6	3 16 8·23
W.—Bathurst	76 5	76 3	75 2	73 1	72 9	71 9	71 3	73 8	75 7	74 3	74 10	74 6	73 10	71 9	74 8	74 8	71 7	3 13 10·58
W.—Hargraves	78 6	78 3	78 6	78 2	78 0	77 8	77 11	78 1	78 6	78 4	78 4	78 3	78 8	78 6	78 8	78 6	78 6	3 18 3·76
W.—Tambaroora	78 8	78 2	78 4	78 1	78 2	78 6	78 6	78 9	78 9	78 9	78 8	78 7	79 1	79 5	79 5	79 4	79 3	3 18 8·76
W.—Mudgee	78 6	77 10	77 5	77 0	78 0	76 5	76 3	77 5	78 5	78 7	78 6	77 10	78 7	78 9	78 7	77 6	78 1	3 17 10·35
W.—Orange	75 2	76 9	77 3	77 2	77 2	74 4	71 7	72 5	70 10	72 3	73 5	73 10	73 6	73 3	71 2	73 5	72 10	3 13 10·82
W.—Stoney Creek	75 6	77 2	77 10	77 3	78 5	79 3	70 3	78 8	78 3	78 10	78 6	78 6	79 0	77 5	64 9	69 7	76 9	3 16 9·23
W.—Forbes	...	...	...	74 9	75 5	75 10	71 10	74 6	74 6	73 7	73 5	74 2	...	...	...	72 7	77 2	3 14 4·09
W.—Wedden	...	...	...	...	...	...	...	...	78 0	...	...	...	...	...	...	...	...	3 18 0
W.—Grenfell and Parkes	...	...	...	...	...	...	...	...	...	80 1	80 3	79 3	78 6	78 9	77 10	76 9	76 11	3 18 6·5
W.—Carcoar and Trunkey	...	...	...	...	...	...	...	...	...	...	...	...	...	71 0	69 0	68 4	65 11	3 8 6·75
S.—Goulburn	77 0	77 1	77 4	77 3	77 8	75 0	77 6	77 5	77 1	77 5	77 6	76 10	79 2	79 5	77 10	74 7	70 9	3 16 10·47
S.—Braidwood	77 1	77 1	77 2	77 10	77 9	77 10	77 10	77 10	77 9	77 6	77 9	77 5	75 9	75 4	74 10	77 2	77 2	3 17 1·47
S.—Tumut	75 8	74 7	73 8	...	78 10	78 6	78 3	77 7	78 3	77 10	77 3	77 10	...	77 9	78 8	75 6	74 2	3 16 11·46
S.—Gundagai	73 7	77 0	72 9	74 5	73 5	77 4	77 9	77 7	76 0	75 10	76 0	77 3	73 10	74 0	70 10	69 11	69 1	3 14 2
S.—Meragie	77 10	79 8	...	...	...	...	...	...	...	...	...	...	...	...	...	...	...	3 18 9
S.—Tumberumba	...	75 11	77 3	77 9	...	...	77 11	78 0	77 11	78 2	78 6	78 11	78 11	70 8	61 3	66 2	66 11	3 14 6·5
S.—Adelong	...	...	74 5	76 0	75 1	76 10	77 8	77 5	78 2	77 5	77 9	77 6	77 2	76 8	74 4	76 4	77 3	3 16 8
S.—Kiandra	...	...	74 7	75 4	75 0	75 3	75 2	75 4	75 8	75 11	75 3	74 10	75 6	74 11	73 3	...	...	3 15 2·76
S.—Cooma	...	...	75 5	75 2	75 4	75 5	74 10	73 6	75 10	79 3	77 5	73 4	76 6	75 7	76 3	75 3	75 11	3 15 9·6
S.—Yass	...	...	78 9	71 2	80 1	78 0	76 5	...	...	82 2	...	...	...	...	...	...	67 7	3 16 3·71
S.—Burrangong	...	...	...	79 7	79 8	79 6	79 1	79 8	79 8	79 2	78 11	78 1	78 9	77 9	75 1	77 3	78 6	3 18 7·42
S.—Nerrigundah	...	...	...	...	81 8	81 9	81 7	81 5	...	...	...	...	...	...	...	...	...	4 1 7·25
S.—Araluen	...	...	...	...	...	...	...	...	...	77 6	78 0	75 9	75 5	76 4	77 4	76 10	77 9	3 17 1·37
N.—Rocky River	78 0	78 9	78 9	78 1	78 0	79 4	79 4	78 10	79 0	78 11	79 2	78 11	78 8	78 10	79 1	78 4	77 6	3 18 8·11
N.—Nundle	75 7	75 8	75 6	75 6	76 1	75 10	73 8	76 2	76 2	76 1	75 11	75 8	75 8	75 7	75 10	75 11	75 5	3 15 7·94
N.—Tamworth	74 7	74 0	75 3	72 9	74 3	72 11	74 6	75 1	75 0	75 7	75 0	74 8	72 6	74 5	67 8	76 4	73 3	3 13 11·82
N.—Timbarra & Tooloom	...	73 5	72 11	72 4	72 6	71 11	72 7	73 2	73 6	...	...	...	...	...	...	...	...	3 12 9·5
N.—Grafton	...	...	...	69 7	68 4	72 4	...	...	...	...	...	...	...	...	...	...	...	3 10 1
N.—Scone	...	...	...	...	69 2	62 9	71 8	76 4	63 1	65 9	65 11	67 3	...	...	63 7	73 7	...	3 7 10·9
N.—Armidale	...	...	...	...	...	...	...	76 6	75 2	76 9	77 7	77 9	78 3	78 6	77 9	75 9	75 2	3 16 11

The following reports from the Wardens and other Officers of the Department will doubtless be read with much interest. Some of these gentlemen were prevented by the shortness of the notice given to them, and the pressure of other engagements, from reporting upon the resources of their districts as fully as under more favourable circumstances they might have done. Some of these reports bear evidence of great care and ability, deserving of recognition :—

Mr. Warden Clarke reports :—

The depression, which is mainly due to the reaction consequent upon the excitement which prevailed in 1872 and 1873, is likely to prove only temporary. At Grenfell the mining population has decreased, having been attracted thence by the discoveries made in the neighbourhood of Parkes and Forbes. An attempt is being made to re-open the "Grenfell Consols Mine," and if successful, it is hoped it will revive the reputation of that field. At Burrangong sluicing operations have been carried on successfully during the rainy season. The absence of a plentiful supply of water prevents large areas from being worked advantageously. At Spring Creek an attempt has been made to work the wet ground, but though the yield of gold was encouraging, and the company displayed much determination and perseverance, the result was not satisfactory to the shareholders. A quartz reef of a very promising character was discovered at Barmedman, in the Bland country, some months since, but considering the richness of the stone the claim-holders have displayed singular apathy in opening up the lode. It is stated that the endeavour to open up the quartz reefs in the Muttama Gold Field will be renewed shortly.

At Adelong the quartz reefs have been worked with considerable vigour; some of the larger companies are erecting machinery for raising and crushing on a large scale; it is anticipated they will; when such machinery is completed, be enabled to work their mines to greater advantage and profit. On the margin of the Adelong Creek near Shepardstown and Grahamstown, some very good alluvial ground has been opened, but the working of this ground is impeded by the position and extent of the land in the bed of the creek, held under the "royalty lease." Some portion of the land held under this lease has recently been worked to advantage. At Upper Adelong a large extent of ground has recently been operated on by means of sluicing, with very satisfactory results. It is surprising that such ground has so long remained unnoticed. The mines at Reedy Flat continue to be highly productive, and the dividends therefrom have been exceptionally large. At the Tumbarumba Gold Field some very extensive and costly works have been constructed, and now promise to reward the enterprise and energy of the gentleman who has constructed them. A tail-race some 400 yards in length has been cut in some places to a depth of 60 feet, in some places it has been tunnelled a considerable distance through the solid bed rock. The race is boxed throughout, and its cost is estimated at about £3,000. By means of this race the lessee will be enabled to ground-sluice some thousands of yards of the Burra Creek and the adjacent flats. It is estimated that the lessee has a field which will afford remunerative employment for the next twelve years. The deposit is from 18 inches to 2 feet in height, and prospects taken from the face at several points gave an average of ½ dwt. to the tin dish. On the Mannus Creek similar works on a still larger scale now approach completion under the superintendence of the same gentleman, and the ground which will be operated upon when the race and other preliminary works are completed is said to be more promising than that at the Burra Creek. These works are without doubt the largest and most important of the kind in the Colony, as they will ensure the development of large areas of ground known to be highly auriferous which have remained unoccupied for many years. The tunnels by means of which the rich deposits supposed to underlie the basaltic formation in the high lands around Tumbarumba were to have been tapped, have not produced such satisfactory results as the enterprise deserved, and work in connection with them has been temporarily suspended. At Ouranie mining operations have not been carried on with that vigour which the known richness of the reef would justify; this may be due to want of mining experience or want of co-operation amongst the miners. At Yarara, near the Ten-mile Creek, Albury Division, some excitement prevails in consequence of the discovery of some gold-bearing quartz veins found in the adjacent slopes, but no opinion can be expressed of the importance of the discovery pending the further development of the veins.

Mr. Warden De Boos reports:—

At the Shoalhaven River, advantage has been taken of the low state of the river to work portions of the river bed. Both Europeans and Chinese are employed upon this work, which is sufficiently remunerative whilst it lasts, though the gold is very fine and requires great care to save. No works of any great extent, however, have been constructed in connection with it. The working of a river bed is at all times a precarious operation, but it is more especially so in the Shoalhaven, which, taking its rise in some of the loftiest ranges on the coast, is peculiarly liable, in the upper part of its course, to floods,—a thunderstorm in the ranges often sending down a fresh heavy enough to demolish the works in the river bed. Thus it can of course only be worked in the driest seasons, and the moment the rain sets in the river bed has to be deserted, the water-wheels removed, if not previously carried away, and the labour of months to be left to be washed out by the stream as if it had never been. The rain, however, fills the races, and sets going another class of work. Sluicing the banks of, and the necks of land which project into the river, is now commenced, and is carried on until the dry season once more cuts off the source of supply. This kind of work gives very payable returns where a good sluice head of water is obtainable, and some of the parties here have gone to very great expense in constructing races. The most noticeable of these is the race cut by the Warri Sluicing Company, which is a well-finished piece of work in every particular. Twenty-three miles of it have been cut, and some few hundred yards have to be added. It will draw its supply of water from the head of Reedy Creek, and will carry about six ground-sluice heads through the hills to the banks of the Shoalhaven, which are to be worked down by hydraulic power. Another race, almost equal in extent to that of the Warri Company, has been cut, and been in use for some time by the Shoalhaven Sluicing Company. This is also worked by means of the hydraulic hose, but at present sluicing parties are almost idle, owing to the small supply of water. Works of the character I have described extend in places down the river almost to Nerriga. In the neighbourhood of the last-named place several promising reefs have been taken up, though at present none of them are being worked. About 150 miners in all are at work here, in a length of 30 miles. I have tried to get the amounts of gold obtained from this quarter, but have found it impossible to do so at present in anything like a reliable form. I am in hopes, however, of being able to make arrangements with the gold-buyers to get correct information on this point in the future.

The Mongarlowe or Little River Gold Field is situated about 8 miles east of Braidwood, and the working is all carried on upon the Mongarlowe, a tributary of the Shoalhaven, and Serjeant's, Fagan's, Bob's, and Nettleton Creeks, which run into it on the north, and the Tantullian and Warrambucca Creeks, which enter it on the south. The auriferous deposits are nearly all found on the north bank of the river, and throughout the whole of the tributary creeks which come into it on that side. The Warrambucca and Tantullian Creeks may be regarded as affluents rather than tributaries, as they are of equal size of stream with the Mongarlowe, and these creeks again have been found to be richest on the northern side, and have turned out a very large amount of gold. The miners here are now only employed in the wet season, when the races are full, the workings in the river-bed being carried on by the Chinese. In the dry season the European miners do a little fossicking amongst the old ground.

Except in the wet seasons, the yield of gold from this field is very small; and, as in the case of the Shoalhaven, there is really no mode of obtaining a reliable return of the actual amounts produced in this locality. By an arrangement with the gold-buyers, I think I shall be enabled to get at something like an approximate return of the yield in future. The workings have been all shallow alluvial sinking, very seldom exceeding 20 feet in depth; and all the auriferous ground, which does not on the average extend for more than some 200 yards from the river, has been pretty well worked out, every creek and gully, and every little break in the ridges, having given up its run of gold, more or less extensive according to the character of the ground. It was originally worked in block claims; but since then very much of it has been ground-sluiced, whilst much of the shallow surfacing on the face of the ridges has also been sluiced, with payable results. It is upon this sluicing that the miners now mainly depend, and they have consequently only the races to look to for a supply of water. Of these ordinary races, cut for the supply of individual miners, over 100 miles are maintained; but besides these, a very fine race has been cut by the Warrambucca Sluicing Company, some 25 miles long, bringing a strong head of water from the Buddawang Range on to the lower part of the Warrambucca Creek. This race is flumed over the Warrambucca Creek, at a height of 10 feet, by something like a quarter of a mile's length of boxes. It is computed that there are about 200 European and 100 Chinese miners working on the Mongarlowe Gold Field. Some reefs have been opened here, with a very good show of stone, some of which was very rich indeed in patches that soon ran out however. The work on these reefs has been discontinued, owing to the parties being unable to keep down the water with the pumping machinery at their disposal. The water-level here is very soon reached, being struck at a depth of 30 feet only; but I am under the impression, from what I have gathered from miners who have been employed on these reefs, that the water is surface soakage only, and not a regular underground water deposit, because at a considerably lower depth no water is found to come in. Thus, in a shaft that was sunk some 100 odd feet, and kept clear of water by pumping, a drive put in some 20 or 30 feet was found to be perfectly free from water, no soakage coming in from the roof of the drive. If this be the case, the water may very easily be kept down, if the parties be disposed to continue their work.

At Jembaicumbene the work is all done by Chinese, and on private property. It is confined entirely to sluicing, but the deposits have been pretty well worked out.

Major's Creek and Bell's Creek come within the same description I have already given of the Mongarlowe Field, being entirely dependent for work upon large supplies of water for sluicing purposes; and as the present dry season has left the races entirely empty, no work, except a little fossicking, and in some cases, where the miners are possessed of means, the breaking down of banks and preparing the earth for sluicing, is being done. Two reefs were being worked here; but one was about to be discontinued so soon as a good patch of stone which had been struck should be worked out, and the other was not expected to continue work much longer. There are not above a hundred miners here, of whom about a third are Chinese; and I could find no means of ascertaining anything about the yield of gold, as it all goes into the Banks at Braidwood.

Coming now to Araluen Valley, the work from which the greatest amount of gold is obtained is the "stripping," carried entirely upon private land. This work, which can only be carried on in a dry season, is of a very extensive character, and necessitates the employment of a large number of hands. The ground being very loose alluvial deposit, completely saturated with water from the underground drainage, can only be worked by stripping and carting away all the surface soil down to the washdirt, often a depth of from 20 to 30 feet. This is done by means of horses and carts, which remove the earth and tip it out upon some spot not required to be worked, or into some "paddock" which has been already worked out. As this work is carried on, the paddock that is being excavated is kept clear of the water from underground soakage by powerful pumping machinery. To give an idea of the work carried on, I may mention that one claim—that of Herbert and party—employs at the present time seventy men and forty-five horses, whilst they have three powerful engines on the ground. One of these is employed in hauling up trucks of washdirt to the sluice-boxes, one was employed in hauling up trucks of refuse to the tips, and the third was engaged in pumping. Owing, however, to the pumping engine getting out of gear, the hauling engine had to be removed and set to work the pumps. The trucks are hauled up by a wire rope on a tramway laid on an incline, and this application of the tramway has been found to secure an immense saving of labour. Only two other extensive claims are at work, and these employ somewhere about fifty men each and thirty-five horses, besides the engine and machinery for pumping. Two or three other claims, working upon the same system, employ a smaller number of hands. These miners, however, who are employed upon private land do not require to have miners' rights, nor do they come under the mining regulations in any way, as even the ground upon which their houses are built, as well as that upon which all the business places have been erected, is private property. Only the extreme upper and lower end of the valley is Government land, and here sluicing operations are principally carried, there being no depth of soil in the creek-bed to favour the deposits of gold, as the bed-rock is washed almost clean by the fierce action of the water confined at these points to narrow gorges. Water, however, is now so scarce that hardly a race is running, and those that have a little do not carry more than half a sluice-head. A little work is being turned at the head of the valley, but very small compared to what it would be were water more plentiful. At the same time, if water became so abundant as to set the races running, the large stripping claims in the valley would be mastered by the water and would have to knock off work. The remarks respecting the head of the valley apply equally to the lower portion, and in fact to the whole extent of the Araluen Creek down to its junction with the Dena River. The two streams when joined seemed to me to carry about two full ground sluice-heads, but certainly not more than three; so that some judgment may be formed of the difficulties in the way of sluicing. The total number of miners, including those on the Lower Araluen, is estimated by persons well acquainted with the locality at about 450; and of these, 250 are engaged in mining upon private property.

Mogo is a very unimportant Gold Field at present, having been completely worked. The deposits were rich for a time, but not very widely spread; and as it depends solely upon sluicing, nothing is now being done there. It is situated in the lower country of the Clyde watershed, and does not contain at present more than some twenty or thirty miners.

Nerrigundah or the Gulph has been a very rich field in its day, but it is now very nearly worked out. The kind of work is precisely the same as that carried on at Major's Creek and Mongarlowe, namely, working the river-bed when water is scarce and sluicing the banks when the races are full. There are about 150 miners on the ground, half of whom are Chinese. Three reefs are being worked, but not extensively, being more for the sake of testing than for anything else, since two of them have a 10-stamper battery in close proximity to them. The gold obtained from here and Mogo is never taken into account in any way. There is no escort, and it all finds its way by steamer through private hands to Sydney. Though of no very great importance as a Gold Field at present, the district is known to be very rich in minerals of various kinds, and it will thus require a good deal of careful attention from the Warden in charge. Besides this, seeing that the Gold Fields of Mogo and Nerrigundah come so close down upon the coast, there is fair reason to believe that other tributaries of the coast rivers may also be found to be auriferous—in fact some auriferous land in this district was reported.

The Delegate Gold Field is distant from Bombala 25 miles, and about 140 from Nerrigundah; this long distance having to be passed over without a single Gold Field intervening, the country also being exceedingly heavy and mountainous. The auriferous ground hitherto worked extends along the banks of the Delegate and the Little Plains Rivers. The latter is more commonly known as the Little River, thus making two Little Rivers in the Braidwood Gold Mining District. It is for this reason that I have throughout used the name Mongarlowe, so that the two places may not be confounded with each other. These rivers and their tributaries have been worked for a distance extending over some 30 miles, from the Little and Big Bogs on the tableland, whence the races are supplied, down to the boundary line which separates this Colony from Victoria. The gold obtained here is of a coarse and nuggety character, very little waterworn, even when procured from the river beds. The working is precisely the same as that I have previously described as followed at Major's Creek and Mongarlowe. The sinking is very shallow, the greatest depth being in the bed of the river, where it never exceeds 20 feet. No reefs have yet been found in this part of the country, although good payable reefs have been opened at Bendoch and Bonang, in the Colony of Victoria, and not more than 10 miles distant from the border. There are about 400 Chinese working on this field, with not more than some twenty European miners. The Neebothery ranges are known to be auriferous, but as yet they have not been opened up by the miners, owing to the difficulty of bringing water on to them. Indications of copper have also been found in these ranges. Here, as in other places I have mentioned, a good deal of gold is obtained, of which no account whatever is received in Sydney, except by the Banks; it all goes down by steamer *viâ* Merimbula, and as it is taken privately by passengers it does not appear upon the steamer's manifests. I was able, however, to trace something like 1,350 ozs., which had been purchased in Bombala by two buyers during the last six months, and I was assured by persons acquainted with the subject that the purchases of these two parties would represent about half the yield, as all the storekeepers buy gold, and very much is taken down by the Chinese. This would give something like 2,700 ozs. for the six months, or 5,400 ozs. for the year, and this, I am told, is actually below the average. The field itself, from the produce of gold, is thus well deserving of notice; but it is all the more so from being on the boundary of the Colony, in close proximity to a payable Victorian Gold Field which draws all its supplies from our territory. In fact, all the traffic with this part of Victoria is carried on through Bombala and Merimbula, and a very extensive trade has sprung up across the boundary.

I may mention that the tedium of a long ride of 110 miles from Bombala to Kiandra was very agreeably broken by an inspection I made of the M'Laughlan River. Immediately on coming down upon it, the whole appearance of the surrounding country struck me very forcibly as giving every indication of an auriferous tract. High basaltic hills border the river on its northern side, many of them being perfectly table-topped, from the covering of trap still remaining unbroken; whilst on the southern side the hills are more denuded, showing the slates outcropping in heavy bars, and their sides covered with quartz and slate debris. I occupied several hours in making an examination of the country, and became firmly convinced of its auriferous, but being on horseback, and consequently without the means of testing the ground, I was unable to prospect it practically. Inquiries which I subsequently made in different quarters brought a confirmation of the opinion I had formed, as I ascertained that gold had been found here in small quantities upon several occasions, but from some unaccountable reason the ground has never been fairly prospected. The river brings down a fine body of water, and there is every facility for the work of the mines, and it is astonishing that it should have remained so long untried. On my next visit to Delegate I shall endeavour to induce some practical miners to try the ground.

Mining at Kiandra is at rather a low ebb at present; but still the few men who are working there are obtaining a fair amount of gold. At the same time, the yield generally is not sufficiently great to induce any large number of European miners to face the inclement weather which almost continuously prevails in this portion of the Colony. It is to this cause alone that I attribute the gradual decrease in the population. Those at present on the ground, miners, as well as storekeepers and others, are men who have been on the spot almost since the first days of the great Kiandra rush; and though they admit that things are now dull, owing to the small population, they still hang on to the place with an expectation amounting almost to certainty of better days to come. The great thing to which all are now anxiously looking to restore the falling fortunes of Kiandra is the extensive works which are now being carried on at New Chum Hill. This hill, like all those on the same side of the river (north) is basaltic, the trap overlying a great depth of alluvial deposits. A very good section showing the formation of the hill is given by a deep cutting at no great distance from the site of the present works. Here the wash-dirt was worked some five or six years ago with very payable results, by a large party of miners, by means of this large open cutting, but the work had to be given up when the cutting, getting deeper into the face of the hill, became too deep to make the work payable. This cutting, which is about 60 feet deep, shows at the top about 20 feet of ordinary alluvion mixed with rounded waterworn boulders. Then come some 10 feet of coarse pipeclay, overlying a 6-feet stratum of what may be termed incipient lignite. Under this lies a stratum of pipeclay some 12 feet thick; and immediately below that is the wash-dirt, varying in depth with the conformation of the bed rock, sometimes running out and sometimes reaching a depth of 12 feet, the average being some 6 or 8 feet. This work has paid in the old time from £10 to £20 per man for sluicing. What I have termed incipient lignite is a bed of vegetable matter, mainly leaves, which, speaking geologically, is a comparatively recent deposit, for some of the blocks which have been exposed to the atmosphere for several years can be opened out with the point of a knife in perfectly single leaves, something after the same manner that a single hop flower can be opened out from the closely pressed mass of flowers on the covering being removed. When fresh from its place of deposit, the block has all the appearance of cannel coal in a first stage of formation, having the peculiar conchoidal fracture of that mineral rather than the longitudinal cleavage of the shales. This being the history of the country, the prospects of remuneration from the wash were so good that a few influential gentlemen in Sydney joined together to test it by systematic working. A tunnel 6 feet high and 7 feet wide has been driven some 350 feet into the hill. A line of tramway has been laid down—the tunnel being wide enough for two lines when the second is required—and the wash will be run out in trucks to be shot into hoppers of a crushing mill of 10 stampers. This mode of treatment is adopted on account of the wash-dirt being of a character to resist the action of sluicing, whilst the richer portion of it—that lying on the bed rock—is agglomerated into a hard and compact cement which nothing but a stamping battery can break up. The tunnel has been driven under the wash, so that all that will be required will be to stope it down into the trucks, and run it away to the machine. By the saving of labour thus effected, and the quantity of wash that can be treated, it is estimated that even as low a yield as 4 dwts. to the load will return a handsome dividend. Hills of a precisely similar formation to that I have described come down as short spurs from the southern face of the main range (the Snowy Mountains) for a length of 30 or 40 miles. They are all basaltic; many of them table-topped, and they have been proved, in several instances, to have the same kind of auriferous wash underlying the pipeclay and reposing on the bed rock. Thus, if the working of this tunnel should prove a success, there is a certainty that, in spite of the inclemency of the district, tunnel claims will be taken up by many others besides the present investors.

The Kiandra alluvial workings extend in a direct line north and south for a distance of some 30 miles; the auriferous deposits being all found in the hills from the southern face of the main range; the Four-mile, Nine-mile, and other diggings being all on the same sequence of hills which forms the north boundary of the Eucumbene River. All the tributaries of this river coming in from the north have also been found to be more or less auriferous. The work is the same as that carried on in all old alluvial diggings, viz., washing the river bed in summer when the water is low, and sluicing the banks and hill sides in the wet season when the races are supplied with water. The greater part of the gold obtained here is coarse and nuggety, that from the Nine-mile being peculiarly so, the nuggets being much mixed with quartz and having a beautifully efflorescent form. They are very little water-worn, and the edges are sometimes so sharp as to have the appearance of being fresh from the reef. From the Four-mile the gold is more scaly, but is heavy and of good quality. In the main river again it is coarse, nuggety, and of a high standard. It is next to impossible to get at anything like an approximation of the amount of gold raised in any given term, as very little of it reaches the Cooma Escort. The great bulk of it is sent away by post to Sydney direct. Some idea of the quantity may however be gathered from the fact that Mr. Lett, J.P., a storekeeper here, assured me that during 1874 he purchased something over 1,500 ozs. There are two other buyers in the town who he says obtained nearly the same quantity, so that we may fairly assume that some 4,000 ozs. have passed through the post, and this amount would be altogether independent of that sent by the Chinese to their countrymen in Sydney. Some guide to the amount of gold posted may be obtained by ascertaining the amount of percentage paid to the postmaster here on the sale of postage stamps. I have heard that that percentage reached £40 last year, owing to the number of stamps required on parcels of gold posted.

Quartz reefs have been opened up here in several localities; but though some have given good returns, they have all been dropped, and none are being worked at present. There is a four-stamper crushing battery at the junction of Bullock Head Creek and the river, but it has lain idle for many months. The number of miners working on this field is computed at 60 Europeans and 100 Chinese, and there is little prospect of any increase in this number unless the tunnelling operations prove successful. Tin has been found some 12 miles distant from Kiandra in the granite country, and several leases have been taken up. The ground has been fairly tested, and lodes of grey streak tin have been opened, but none of the ground is being worked at present. Copper lodes have been opened at Yarrangobilly, better known as Lob's Hole, about 15 miles from Kiandra, on the road to Tumberumba, by the north and the south Yarrangobilly Copper Mining Companies. I inspected the lodes of the latter company, and found some very good specimens of carbonates, red oxide, and yellow sulphurets, but the terribly broken country in which the lodes are situated is very much against their fair development.

Making my way through the immense mountain ranges that lie between Kiandra and Tumberumba, I reached the new and old Maragle diggings, situated about 20 miles east of the latter locality. The alluvial which at one time gave very rich returns is now worked out, and operations are confined exclusively to sluicing. Some twenty or thirty men find employment here.

Tumberumba Creek and gold mining district are also essentially sluicing localities, the miners being entirely dependent for returns upon the supply of water to their races. The men are scattered over a very large extent of country, almost every tributary of the creek being worked by races. In fact the whole length of the Tumut with its many feeders gives employment to the sluicers, the line of auriferous country extending all the way between Tumberumba and Tumut. It is estimated that between 300 and 400 men find employment in this line of country, every little gully furnished with a stream of water having, according to its extent, its four, six, eight, or nine miners at work in it. From here the yield of gold is pretty correctly shown by the Escort returns, as the metal nearly all passes through the hands of the Banks and goes down by Escort, and there would be little for me to say about the spot beyond what I have already written respecting the other sluicing grounds, were it not for the extensive operations which are being carried on by Mr. Gitchell at the Burra and the Mannus Creeks. Both these creeks are being worked upon the same principle, and it will therefore be necessary to describe one only, and as the works at the Burra have been completed, and are now in full operation, I shall select that. Leases have been taken up of 1,500 yards of the creek bed. To work these an extensive tail-race had to be cut, at an expense of £3,400, before a shilling of return could be obtained from the ground. The race is 400 yards long, one-third of the distance being tunnelled, as the granite was found to be so hard that this was considered the cheapest mode of proceeding. The mouth of the tunnel is guarded by a valve or flood-gate, so that in the event of a flood so heavy as to be likely to carry off the gold the opening can be closed, and the waters thrown off into the original creek bed. The bed rock lies some 16 feet below the present surface of the creek, and the race has been cut down 14 feet into the bed rock, giving a fall of 30 feet from the surface to the line of the race. It is boxed throughout, the boxes being 4 feet wide by 2 feet deep, and laid with a fall of 1½ in 12. The deposit in the creek consists of rich black alluvium intermixed with boulders overlying the wash-dirt, which runs from 6 inches to 2 feet deep. On the surface is a heavy coating of long, tufty, sedgy grass, the roots of which hold the soil very firmly together. When this is once cut through, the waters of the creek tear away the lighter deposits below, and carry them down into the sluice-boxes, where the heavy stream of water soon carries away every particle of alluvial matter. The advantage of this kind of work must be manifest, as the race once formed only a comparatively small amount of labour is required for carrying on the work; and with half a dozen men an acre a fortnight may be completely sluiced away. The work must therefore be profitable even though the yield of gold was comparatively small; as it happens however the ground is particularly rich. Several dishes of wash-dirt, taken at random from the undisturbed portion lying on the rock were panned out in my presence, through the courtesy of Mr. Gitchell, the managing director of the work, and in each instance the yield was somewhere about half a pennyweight of gold. The same gentleman also informed me that some time back he washed out about an acre of the creek bed, and that the clearing up of the boxes gave him over 200 ounces of gold and about a ton of grey tin ore. He expects that will be about the average rate per acre. If it be so this speculation will be a fine one indeed, and will be almost sure to lead to the formation of copartneries of a similar kind to work localities of a like description. There are hundreds of such places, more particularly in the Southern District, to be found, where miners have been unable to sink on account of the water, and where there is plenty of fall for the construction of a tail race. In all these places this system of work would be quite as applicable, and I believe the returns would be equally as good. The works at the Mannus Creek are even more extensive than those at the Burra, and before they are completed will cost over £6,000. It is enterprise of this kind to which every legitimate encouragement should be given, as it is that only which is required to make our Gold Fields as productive as they should be. No quartz-reefing is being carried on in the district.

Work in the Tumut District is of precisely the same character as in the Tumberumba Division, the work being all race-cutting and ground-sluicing, whilst one local company, the Great Britain, has commenced the cutting of an extensive tail-race to work their ground on the same principle as the Burra. There may probably be some 200 or 300 miners in this Division, but from their being so scattered about in every gully and watercourse it is next to impossible to make any correct estimate of their number.

Adelong has been the steadiest reefing district in New South Wales, having kept up the supply of gold pretty regularly for the last ten or twelve years. The principal line of reef now working is the Victoria. Upon this four companies are steadily at work, the chief of these being the Williams and the North Williams. The outcrop of this reef was found very near the crest of the lofty hill which overhangs the town of Adelong on the north, and it has been sunk upon to a depth of 530 feet. No water has been encountered beyond ordinary surface soakage that can be kept down easily without pumping. Good stone is now being obtained at this depth, and I saw some good stone sent up from the bottom of the shaft of the Williams Company that would probably go five or six ounces to the ton. I was however more especially pleased at seeing in the North Williams the first approach to a systematic mode of working the reefs. In this reef the stone is stoped into trucks, the trucks are run along the galleries on to cages, are sent up without shifting the stone, are run from the cage on to tramways, from which they tip their loads into the heaps for which they are intended. The hill on which the shafts are situated is very precipitous, and the crushing machines being worked by water power, lie at the foot of the hill in close proximity to the creek. There is thus some difficulty in making a tramway from the shafts to the mill, but in the case of the Williams Company this has to some extent been met by the construction of a tramway from the mill to the more precipitous part of the hill, down which it is proposed to deliver the stone by a shoot. The North Williams Company have taken up a dam and a race site, and intend shortly to erect crushing machinery of their own in such a position as to allow of a tram to be run to it. The eight-hour system has been introduced here, and the men consequently work three shifts in the twenty-four hours, the work being continuous night and day. The yield from the stone throughout, since these reefs were first opened, has averaged a very little under 4 ozs. to the ton, though at present the stone brought to grass barely comes up to the average. A gentleman who is very intimately connected with gold-mining in the district, and has an interest in looking after the pay sheets of the various companies, has informed me that from the number of men employed, if the gold sent by Escort from Adelong falls below 700 ozs., mining in that district does not pay, that being about the amount required to cover wages expenses. I believe my informant to be perfectly trustworthy, but it would require a longer experience of the field than I possess to pass any opinion of my own upon the subject. The main alluvial workings upon the Adelong Creek are now carried on at Shepherd's Town and Graham's Town, the former three and the latter five miles from Adelong, on the road to Gundagai. The Graham's Town workings are turning out a very good amount of gold. There are some ten claims taken up upon a rich patch of ground which was opened some few months back upon a bank of alluvion deposited at the foot of one of the spurs coming down from the range which forms the eastern watershed of the creek, and evidently covering an old creek bed. There are from 6 to 8 feet of wash-dirt, some of which has given as much as 2 ozs. to the ton, the average being however something less than 1 oz. to the load. The upper ground being pretty firm, the wash is driven out, being raised by water power, whilst the pumping necessary to keep the ground clear of water is done by steam power, there being three portable engines employed upon this work. The bulk of alluvial gold going to Sydney from Adelong comes at present from this locality. There are not more than about 200 European miners working in this Division of the District, with some fifty or sixty Chinese scattered about amongst the old workings.

In the vicinity of Gundagai some few quartz claims are being worked, but they were not sufficiently extensive or important to induce me to visit them.

Passing through Gundaroo on my way back to Braidwood, I found upon inquiry that upon the reefs opened in this District work had been discontinued for some months past; and that with regard to the alluvial ground, although gold in small quantities had been found by a few miners in different spots amongst the ranges, yet nothing like a sufficient deposit had been struck to induce any large influx of miners.

Mr. Warden Johnson reports:—

I regret that I cannot report as favourably as I could wish upon the Districts under my charge, namely, the Bathurst Mining District and the Tambaroora and Turon Mining District. The alluvial and river-bed workings, once so extensive and remunerative, are apparently in a great measure exhausted. Quartz-mining is, therefore, the source from which the supply of gold must in the future be almost entirely expected. This is a branch of gold-mining of a nature demanding capital—now, on account of the serious losses during the late mining mania, so difficult to obtain for the prosecution of even legitimate operations with every prospect of success. The principal mining centre is Hill End, where mining operations are still being pushed forward on the Hawkin's Hill line, with an energy and perseverance that deserves, if it has not yet commanded success. A system of tribute, remunerative both to the owner of auriferous properties and to the tributer, has been established on this Gold Field on several of the quartz mines which had been opened up but could not be profitably worked by daily labour under expensive supervision. This is the most hopeful feature of mining in the district. Fair and even large yields are still in a few instances obtained, but the majority of the mines "floated" at Hill End for large sums are still unproductive—the problem as to the continuance of rich deposits at a depth below 500 feet remaining unsolved. But as sinking is still being carried on in some of the claims which are possessed of sufficient capital to continue operations for that purpose, it is to be hoped that another year will not elapse without setting this important question at rest. Not being aware until a few days since that a report was required from me, I had not prepared myself with that detailed information in regard to the mining operations of my districts which would be necessary to complete this report, and which I cannot now obtain within the time allowed, I must plead as an excuse for its general and somewhat desultory character. Copper mines are scattered over the Bathurst District, from Wiseman's Creek near the upper waters of the Macquarie to the vicinity of Cowra on the Lachlan River. But here again I am not possessed of that statistical information or description in detail with regard to these mines which would enable me to report authoritatively upon their extent or productiveness. I may state, however, that although no great financial success has yet attended the working of these mines, a large quantity of fine copper has been smelted on the spot, and that the existence of a rich belt of cupriferous country only partially developed has been conclusively proved to exist in the Bathurst mining district, and with the advance of the railway and the consequent increased facility for the carriage of ore and coal to and from the coal fields at Bowenfels must evidently prove a material source of national wealth.

Mr. Warden Dalton reports:—

Commencing with the Billabong Gold Field, which is now virtually an extension of the Lachlan Gold Fields, I beg to refer to my annual report for the year 1873, and the description therein contained of such portion of the field as is situated on the north bank of the Goobang Creek, and to which mining operations had, up to that date, been confined. On following the range of hills therein described in a south-westerly direction from Currajong, it becomes apparent that to the upheaval of granitic masses and the welling up of trappean products in the larger fissures thereby created, is due the mineralization of the adjacent sedimentary deposits, their metamorphosed or altered conditions, and the subsequent filling of minor cavities and fractures, either by infiltration or hydrothermal action with silica and other mineral matter capable of either solution or sublimation. Evidence of this may be traced on the surface in fragments of altered sedimentary and trap formations mingled with ferruginous quartzites in the vicinity of a crest of granite boulders, with the outcrop of a reef or vein of quartz trending across a hill at no great distance.

Although the surface of many of the hills presents sufficient indications that granitic or other transmuting agencies are not far distant, these latter have not been denuded. In such localities the operations of the miner frequently disclose trap dykes filling short irregular fissures, and slates, and interstratified sandstone, tilted, contorted, altered, riven, and fractured in every imaginable direction. These broken and disturbed strata are often silicified and reconsolidated by threads and veins of a metalliferous quartz or calcspar, and sometimes these threads and veins are highly auriferous in small patches. Occasionally they occur so frequently as to give the stratum the character of a lode, but more generally such formations only tempt the miner to prosecute a search for a reef that he is not destined to find; hence we hear so much about leaders and hopes to strike the reef which are never realized. It is in similar formations, similarly acted upon by eruptive trap dykes, where the iron has been converted into a brown oxide, and the silica and lime removed in solution, and the strata decomposed and converted into a red-brown striated clay, that small nests of golden nuggets are found *in situ* where there is neither gravel nor watercourse, and perhaps a single find is the sole return for the labour of many months and a large expenditure.

When these interstratified sandstones occur and there is no evidence of cleavage or jointing, the breaking up has been most irregular: and as the resistance to the upheaving masses has been overcome by a general fracture of the superincumbent deposits, it is improbable that any well defined fissures will have been formed; the result has been the narrow veins and threads of quartz that so abundantly interlace the older stratified deposits.

I have been somewhat particular in my description of this peculiarity of this section of the district; many have foundered upon the formation I have described, and few understand it; more than one well organized company has attempted to sink through such a stratum tilted to an angle of 80 degrees, in the hope of finding an El dorado at the bottom.

On several leads discovered and worked during the past year it has been proved that the deepest ground is not auriferous; shafts have been sunk to depths varying from 110 to 160 feet and bottomed upon a wash dirt that differed in no respect from that obtained in adjoining claims at depths of from 80 to 90 feet, except that the former did not contain any gold, while the latter yielded from 10 to 15 pennyweights per load; these watercourses run in a parallel direction, and are not 100 yards distant from each other. (*Vide* Section A annexed.)

A band of argillaceous limestone of an unascertained depth and width, in which no fossils have been as yet discovered, extends from the bank of the Lachlan River due north, and intersects the Lachlan and Billabong Gold Fields near the centre; cropping up on the limestone ridges about one mile to the west of Forbes, it again appears in the midst of surfacing at the head of one of the main tributaries of the Bushman's Lead, reappears near the Dayspring reef, where its eastern margin is intersected by trappean dykes, and again comes to the surface to the northward of the No Mistake lead, whence it passes onward through untried country.

The most productive auriferous leads both on the Lachlan and Billabong Gold Fields have been opened on the flanks of the belt of limestone referred to, and the leads now being worked to the south of the Goobang Creek on the Reserved Gold Field of the 22nd of June, 1874, follow the same course and are similarly situated with respect to it. Nuggets of gold weighing from 2 to 9 ounces are frequently obtained from these leads enveloped in what appears to be a decomposed silicate of lime.

In the centre of the surfacing previously mentioned a cavernous opening was discovered in the limestone, about 60 feet in width and 120 feet in length, the walls on all sides being perpendicular. This cavity was filled with the ordinary wash dirt which had been drifted into it, and has been worked from the surface to a depth of 50 feet with payable results; as yet no bottom or horizontal opening has been reached. This cavern must at some former period have received the drainage of a wide valley; the gravel and debris forming the surrounding surfacing appears to have been a lacustrine deposit. This limestone may have exercised a strong influence over the trappean formations reposing upon its flanks, and to some extent it may be acceptable as a guide to new discoveries.

Large quartz reefs may be occasionally observed where there has been but little disturbance of the schists, but as a rule they are not metalliferous.

Thus far I have endeavoured to convey some idea of the lithological formation of this Division of the District, so that the progress made in its development as a gold-mining country may be the more clearly understood.

At the commencement of the year, the gold-mining population having largely increased, and the Bushman's and Welcome Leads with their tributaries being nearly exhausted, numerous prospecting parties spread themselves along the range of hills already described. The result was the discovery on the south-western slopes of the London, the Ben Nevis, and M'Guiggan's, with many other auriferous leads of minor importance. These leads, as they descend to the valleys and flats, become deep, narrow, and tortuous in their course; their source generally is in surfacing or shallow ground on the slopes, and it required much labour and perseverance to trace them into the deep ground where some were lost and others followed to the Goobang Creek. (*Vide plan of leads annexed.*) Of these M'Guiggan's gave promise of being the most productive, and at an early period formed the centre of attraction, and 5,000 persons were speedily concentrated upon it. The first mile below the Prospector's having proved to be less than the frontage depth, was allotted in block claims, the remaining half-mile extending to the margin of the creek being the required depth, was occupied under the frontage system. Whatever surface indications might be found upon the hills and ridges, there were none to be observed on the valleys or flats; no watercourses, no outcrop of rock, nothing but an occasional patch of swamp or shallow gilguey holes; and it was no uncommon thing for a party to sink and drive 1,000 feet before they struck the lead.

As these leads became gradually developed and were traced to the margin of the creek, and it was found to be above the course of the ancient streams and to contain no auriferous deposit, it was clear that the present channel of drainage had been formed subsequent to the denudation of the hills and high lands on its north bank, and that the bed of an older stream must lie further to the southward and be covered by the alluvial deposits that stretch towards the Lachlan River. Impressed with this fact, adventurous prospecting parties crossed the stream, and scattering over the scrubs and plains extended their operations to within a few miles of Forbes; they were followed by hundreds of shepherds, who occupied long lines of imaginary leads in anticipation of the red flag. Land speculators also followed like sharks in their wake, and suddenly discovered that a waterless country that had been permitted to lie unoccupied by permanent settlers up to the present time was the only place where they could select homesteads. At the same time M'Guiggan's Lead was extended and occupied for a mile and a half due south on the southern side of the creek, where a continuation of the lead was sought for with unabated industry.

As the prospectors proceeded with their labours they discovered that the depth varied, and that long ridges, some containing quartz reefs, lay concealed beneath a level surface. The miners had yet to learn that the deepest ground was not the most auriferous, and every new ridge, or as they designated it, mullock bank, created a demand for a new line. The swinging of base lines was incessant, and no little confusion and litigation was the result. In fact, the operations of the miners on these wide levels appeared more like the evolutions of an army in the field than anything else—marking claims had resolved itself into a game of chess. At last, early in July, in the centre of the area reserved from conditional purchase on the 22nd of June last, near an isolated hill of quartzite that rises in the plain half way between

Forbes and Parkes, Murray and party hoisted their flag and reported payable gold, when the Tichbourne Lead was declared a mile east, and a mile west of the prospecting area. Subsequently the claims or areas on the eastern line were proved not to be of the depth required, when the frontage system on that portion of the lead was annulled. This lead appears to have its source in the hill referred to.

The red flag next waved over M'Guiggan's South, then over the Wapping Butcher, a rich lead from the same source as the Tichbourne, but having a northerly course, it subsequently fluttered over the Fairy Lead, and now proclaims the Fulton to be payable. These five leads are to the southward of the Goobang Creek, and situated within the area reserved from conditional purchase; they seem to be tributary to a main channel as yet undiscovered.

Prospecting parties to the northward of Parkes have not been successful.

Notwithstanding the persevering industry of numerous parties of prospectors, the labour of a very large number has produced no return; much of this may be attributed to their want of a knowledge of the district, and more to a deficiency of scientific information. As a rule, miners are guided almost entirely by practical experience, and as a sequence, when they get into a new country and amongst formations to which they have been unaccustomed they have everything to learn, often at the cost of an enormous waste of labour.

About one-fourth of the labour of the entire working population during the past year has been wasted in walking to and returning from distant undeclared leads, by those who are more disposed to trust to the labour and skill of others than their own. As the shepherding system is now abolished, it is unnecessary to do more than record the fact.

During the year 1874 the population fluctuated between 8,000 and 10,000. As discoveries were made to the south-west a large proportion moved in that direction, new villages sprang up as if by magic and others as quickly disappeared, the old workings were also to some considerable extent deserted; they will still, however, for a long period afford employment to those who are satisfied with a sustenance and cannot do better elsewhere.

The alluvial leads discovered and worked up to the close of 1873 are nearly exhausted; their total yield was about 50,000 ounces of gold. Their present state is as follows:—

The Bushman's Lead, with its numerous tributaries, has been partially re-occupied as abandoned ground throughout its entire length by parties who are blocking out poor patches left unworked by the original occupants; a few of these realize average wages, the remainder obtain from 3 to 4 dwts. per load. The wash-dirt varies in width from 12 to 60 feet, in thickness from 12 to 24 inches, and is frequently split into several distinct runs by banks of mullock.

The Great Northern is a supposed extension of the Bushman's, at a depth of 93 feet. It crosses the Goobang Creek, and enters a small circular lagoon about 200 yards in diameter, which is filled with washdirt to the thickness of from 3 to 4 feet. This ground, held in block claims, yields from 7 to 13 dwts. of gold per load, and is worked without difficulty. From the lagoon the Great Northern extends northward in a direction parallel to the course of the creek, but up stream. It was held under the frontage system for one mile from the block claims, but did not prove payable, and was abandoned on the discovery of the leads to the south-west. A few claims have been recently re-occupied, and some good prospects obtained. There is much doubt as to the course of the outlet from the lagoon.

The Main Welcome, 3 miles to the westward of the Bushman's, is now nearly deserted; a few claims at its head are still occupied, together with three at the lower end, where the lead pursues a course parallel to the creek, but up stream. Here, at a depth of 120 feet, there is an inexhaustible supply of wash that will yield from 2 to 3 dwts. per load. Such wash can be procured from any of the abandoned claims on this portion of the lead until further mining is prevented by an influx of water.

Paddy's Flat, worked out as a lead; occasional patches of poor surfacing are found at its head.

Victoria Lead, having its source in a cluster of reefs on the western slope from Paddy's Flat, worked out during 1873. Recently a new run has been discovered one mile in length, depth 12 to 14 feet, yield from 3 to 7 dwts. of gold per load, narrow and easily worked.

Reid's Gully, a tortuous narrow lead discovered about the close of 1873, parallel to the No Mistake, a high ridge intervening, after a course of a mile and a half, forms a junction with the latter, depth 40 feet at the head, gradually deepening in its onward course to 97 feet. Some of the claims on the upper portion of this lead yielded from 1 oz. to 15 dwts. of gold per load,—the majority were payable to the junction. Below that point, during the year 1874, three prospecting areas were occupied, and the intervening spaces held by shepherds. At a depth of 115 feet payable prospects were obtained off a decomposed diorite; a washing followed with a yield of 2 dwts. 12 grains per load, upon which the line was abandoned up to the junction. The block claims higher up the lead are still productive; thickness of wash, variable. The No Mistake Lead is still occupied by a few parties who are content with a small but certain return for their labour.

Richardson's Lead (length, 1½ mile) is a tributary to the Welcome; depth, 30 to 60 feet; thickness of wash, 9 to 18 inches. Several nuggets have been obtained from this lead, weighing from 3 to 9 ozs. Quartz reefs abound in the ridges on its southern flank. Some of the claims about the centre of the lead produced from 15 to 8 dwts. per load, the yield of the remainder was barely payable. A few claims on this lead are still occupied and worked.

Sparling's Camp, not proving payable, was abandoned early in the year. The area selected by the prospecting party was in the centre of a wide flat, where, at a depth of from 80 to 100 feet, a good wash was obtained, that appeared to be spread in every direction over the bottom; this wash contained from 2 to 4 dwts. of gold per load. That portion of the Gold Field has not been sufficiently prospected.

During the year 1874 several minor leads were, after much prospecting, discovered in the valleys on the north bank of the Goobang, all trending towards that stream; they were from ½ a mile to ¾ of a mile in length. A few claims about the centre of most of these leads were barely payable; the depth ranged from 30 to 90 feet, and the wash from 9 to 18 inches in thickness. They are as follows:—The Band of Hope, Donald's Gully, Sydney Clinker, All Nations, Gilgney's, The Union, The Tearaway, The Caledonia, The Little Welcome. In addition to these were several large areas of surfacing, all of a similar character.

The undermentioned leads, more important in their character, situated on the north bank, and having their source amidst quartz reefs and transmuted formations, descend to the margin of the stream:—

The Nibler's Lead.—The prospecting claim yielded 1 oz. of gold per load; depth, 25 to 30 feet; wash, 12 inches thick. On the remainder of the lead the claims produced from 3 to 5 dwts. per load. Six parties now working; greater part of the remaining ground worked out.

The Growlers' Lead.—Narrow and difficult to trace; 2 claims yielded 18 dwts. per load; remainder, wages; depth, 14 feet.

The Well-tried Lead has its source, in surfacing, near the head of M'Guiggan's Lead. Some of the central claims produced 14 dwts. per load, the remainder averaged 6 dwts. Twenty-six claims now occupied; remainder worked out; depth, from 6 to 35 feet; wash, from 12 to 24 inches—narrow.

The Frenchman's Lead, about 1 mile in length, is situated on the slopes of a ridge between Paddy's Flat and the Victoria Lead; in some places it expands and becomes surfacing; depth, 12 to 14 feet; thickness of wash, 3 feet; yield ranges between 3 and 14 dwts. per load—nearly worked out.

The four preceding leads were discovered in the early part of 1874.

I now come to M'Guiggan's Lead, which, as regards the north side of the creek, is the most important discovery during 1874. Amongst the same hills from which on opposite slopes the Well-tried and other leads descend, two narrow leads commencing in shallow ground approach each other from the north-east, and uniting at the head of a wide gully pursue a winding course a mile and a half in length to the Goobang Creek. Here, about the 1st of March last, M'Guiggan and party took up an ordinary block claim, and at a depth of 48 feet reached payable gold, three dishes off the bottom yielding 1 dwt. The ground was speedily occupied under the frontage system, but not proving to be the required depth was ultimately held in block claims for 1 mile north and south of the Prospectors. The lead, although deepening in its southern course from 48 to 96 feet, and when held in frontage areas from 103 feet to 114 feet and 115 feet, on the margin of the creek, was very productive, 1 or 2 ounces of gold per load being no uncommon return from the puddling machines. The frontage claims proved nearly as productive as usual on this field; the centre of the lead was richest in gold. Many of the upper claims have been worked out, but there are still about thirty-five occupied on the block system and seven on the frontage. The bottom is a pipeclay, with occasional patches of compact clay slate, sandstone, and decomposed diorite. The connection of the M'Guiggan's Lead north and M'Guiggan's south, on the opposite side of the Goobang Creek has not yet been clearly traced.

Three miles from M'Guiggan's on the opposite side of the high lands to the the north-west, amidst a dense pine scrub, is the London Lead, discovered in January 1874. Next in importance to the former, this lead in common with those previously noticed proved to be poor at its source; but from No. 5 to No. 19 it gradually improved, and in the deep ground beyond No. 22 the course of the true lead has not yet been discovered, although it has been sought for with persevering industry along a line of protected areas 2 miles in length, which have been frequently abandoned and as often re-occupied. On that line gold has been frequently obtained at long intervals, but not in payable quantities, and never in the deepest ground. The depth of the lead was from 27 feet at its head to 95 feet in No. 20; to that extent it was held in block claims, and thence onward under the frontage system. The richest portion of the lead has proved to be the lower blocks, being from No. 8 to No. 19 inclusive, at a depth of from 60 to 80 feet. A deeper channel runs parallel to the lead, but the wash therein is not auriferous. In several of the claims a consolidated drift or cement occurs; this has been proved by a recent crushing to be as rich in gold as the ordinary wash-dirt, and consequently many claims abandoned as worked out have been re-occupied. There were about 500 men employed sinking and shepherding upon the lead during the first half of the year, and the block claims from the source to No. 22 are still profitably worked.

The Little Wonder, an eastern tributary to the London, crossing the village of that name, forms a junction with the main lead at No. 15; depth, 70 to 90 feet; tortuous and narrow, but rich; contains about six claims.

The Ben Nevis Lead, about a mile distant from the London, pursues a parallel course, and a description of it would be but a repetition of that of the latter. with the exception that it has been much less productive; it has been worked for 2 miles in length and rich prospects occasionally obtained. Small washings from this lead have yielded 10 dwts. of gold per load, but the average would not exceed 5 dwts; the whole will be worked hereafter when water is available, the nearest being now about 3 miles distant. Depth of sinking, from 35 feet at the source to 120 feet at the southern extremity; thickness of wash-dirt from 9 to 24 inches; average, 15 inches. The most valuable claims are those about the middle of the lead, the deepest ground being the least productive.

As a rule the wash-dirt on the Billabong Gold Field is from 12 inches to 24 inches in thickness; there are portions of each lead where it is thicker, but these are of rare occurrence and continue only for short distances.

It is difficult to obtain information as to the value of claims on the various leads, but I have been enabled to ascertained from those through whose hands the gold has passed the undermentioned facts:—

From the Great Northern Blocks four men obtained 800 ozs. of gold; six men, from a claim in which there is still 18 months work, 600 ounces; four men for 6 months labour, 450 ozs.; and four men, in return for 4 months work, 350 ozs.

From M'Guiggan's lead, four men obtained, as the produce of two washings, 650 ozs., and have still two years' work in the same claim; six men, as the result of two washings, obtained 656 ozs.; four men, claim worked out, 500 ozs.; four men, claim not worked out, 500 ozs.; six men, the proceeds of one washing, 400 ozs.; and from two men, still working in the same claim, 300 ozs.

At the London Lead four men, from a claim that they are still working, have procured 400 ozs.; six men, as the produce of one washing, 200 ozs.; four men, the produce of one washing, 145 ozs.; four men, one washing, 190 ozs.; four men, one washing, 160 ozs.

From a surfacing party at the Frenchman's Lead, 120, 100, and 150 ozs., the produce of three washings.

From Richardson's Lead, three parties, each at one washing, have had a return; the first of 90, the second of 75, and the third of 80 ozs. of gold.

From claim-holders on M'Guiggan's, one establishment alone informs me that they have purchased about 9,000 ozs. of gold during the last six months, the whole of which is alluvial.

These facts will prove the value of the Billabong as an alluvial Gold Field.

Having described the alluvial gold workings on the north side of the Goobang Creek, the next in order of discovery are those on the south side of the same stream, in the centre of the area reserved from conditional purchase on the 22nd of June last, and upon which three-fifths of the mining population of the Gold Field is now located, either as the occupants of payable claims, or as prospectors.

The first lead in the order of priority of occupation is M'Guiggan's south, extending towards Forbes, one and a half miles. This lead on a north and south line was occupied under the frontage system early in June last. At that period the miners sought exclusively for the deepest ground, and being unacquainted with the former position in descending order of the auriferous formations on the high lands, or with their thickness, or the period of their disintegration, were still more ignorant of the place occupied by the debris of those formations in the alluvium of the plains and valleys, or at what level it was deposited. Shaft after shaft was sunk and the result was ever the same,—wash in abundance, and gold, but not payable. The miners burrowed through the deep ground until their means and energies were alike exhausted, and many had abandoned their areas when the holders of a claim on the line a mile from the creek, at a depth of 157 feet, bottomed upon good wash in a well defined channel containing coarse gold; a subsequent washing produced 18 dwts. per load. Thus stimulated, the miners on the lead made further efforts in the deep ground, but still the payable claim stood alone. At last a shaft, sunk at random 100 yards to the east of the old line, bottomed at a less depth by about 15 feet on the eastern side of a reef or ridge, and struck the long-sought lead; and now all the abandoned claims are reoccupied, and the red flag flutters over nearly every shaft for upwards of half a mile along the line, and others still further in advance. The depth of the original line of shafts is from 120 to 157 feet, deepening as it recedes from the present channel of drainage. The course of the lead is parallel to that line, but not so deep by from 15 to 20 feet.

The shaft 157 feet deep is supposed to be on a distinct lead, as both the wash and gold differs from that of M'Guiggan's south.

In an easterly direction, three miles distant from M'Guiggan's south, is the head of the Tichbourne Lead, having an east and west course, the eastern extension from the Prospectors is now held in block claims, and the western line, under the frontage system, extends to the eastern boundaries of the protected areas on M'Guiggan's south, which No. 19 adjoins. All the claims to the eastward of the Prospectors are payable, with one exception, and several are rich. Of the nineteen frontage holdings to the westward six have struck the lead, and are on payable gold, the yield ranging from 13 dwts. to 6 dwts. per load. Two have not yet proved their wash, three have not yet traced the lead, and eight have not bottomed. The depth of the frontage areas is 106 feet in No. 1 west, deepening in their progressive course to 130 feet at No. 11 west; beyond that it has not been ascertained, unless the 157-foot shaft on M'Guiggans proves to be on a continuation of the Tichbourne, which, if extended beyond No. 19, must cross the former. The block claims to the east of the Prospectors range from 72 to 90 feet in depth, and the wash varies in thickness from 12 inches to 3 feet. 100, 120, 150, and 250 ounces of gold were obtained from four trial washings from four of these claims.

The Wapping Butcher Lead, at a right angle with the Tichbourne, pursues a course to the northward along the crest of the shallow ground. At a depth of from 45 to 55 feet a coarse wash is found, from 2 to 5 feet thick. A trial washing from one of the claims of forty-six loads yielded 109 ounces of gold. From others, nuggets of gold, from 1 to 3 ounces, have been procured, and from 3 to 5 ounces have been taken off the bottom of a shaft. This lead has been traced to the extent of about three-quarters of a mile. Nearly all the claims within that distance are declared to be payable, but, as the discovery has been recent, and the claims not much more than bottomed, any estimate of the valne of the lead would be premature.

The Fairy lead, near the head of the Wapping Butcher, a recent discovery, also contains some payable claims. Here also nuggets of gold have been picked out of the wash. The depth is about 50 feet

The Fulton Lead, nearer to the creek than any of the preceding workings, is about 1 mile in length. There are about seventeen parties sinking upon it. Trial washings have yielded 8 dwts. per load. It is not yet sufficiently proved; however it has been reported as payable. Depth, from 60 to 75 feet.

At the Bald Hills, near the south-west corner of this Gold Field, and supposed to be within the boundary of the reserve from conditional purchase of the 22nd of June last, prospecting has been carried on for the last six months without intermission by Mathieson and party. I am informed that they have obtained a payable drift at a depth of 170 feet. I have received no official report of the fact. The locality is within 5 miles of Forbes.

There are several well-organized parties of prospectors now scattered over that portion of the Gold Field situated between Forbes and Parkes; they all report having obtained gold, but not as yet in payable quantities. The formation is everywhere the same as regards its general character.

During the past year 126,200 loads of auriferous drift have been passed through the puddling machines, at an average of 3s. 3d. per load, or a total cost of £20,507 10s. The average yield was 7 dwts. 14 grains per load, and the total produce of the alluvial mines, 47,868 ozs. 2 dwts. 22 grs.; the increase being considerably in excess of 100 per centum on the production of 1873.

Alluvial mining operations within the Billabong Gold Field have been extended over 140 square miles of country, exclusive of prospecting areas occupied beyond those limits. Although it is probable that each of the alluvial leads will be exhausted within two years of their discovery, the auriferous formations of this portion of the district cover such an extensive area, that I anticipate the opening of a succession of leads as the miners acquire a more perfect knowledge of the rocks of the locality, the conditions under which they are filled with mineral veins, disintegrated and decomposed, of the original channels of drainage and how they have been silted up, and the manner in which the ancient plains and valleys have been buried beneath an alluvial deposit from 50 to 150 feet in depth. The plains of the Lachlan are alluvial Gold Fields wherever gold-bearing reefs or veins are found, however poor, in their vicinity to occupy elevated positions, and where on those high lands, rocks of the Silurian or Devonian era exhibit traces of transmutation or disturbance by eruptive trappean products. Long ranges and ridges of such country flank all the great plains of the Lachlan, more particularly to the north-west.

Quartz reefs and veins.—But six claims or holdings upon quartz reefs or veinstones have been worked by the occupants during the year, one continuously, and five for short and irregular periods; of these latter but one party proved the value of the stone by crushing 100 tons, which yielded 1 oz. 7 dwts. per ton. The alluvial mining that has been prosecuted with so much energy has proved the existence of many auriferous reefs or veins, the exact position of which has not been ascertained. Many of the quartz reefs of the district might be profitably worked by co-operative companies.

Of the eight mining companies formed upon this Gold Field, whose properties at no distant period represented in the aggregate £283,000, six have been inactive during the entire year. Two have disposed of their plant, and one of their entire property on the Gold Field.

The Dayspring Gold Mining Company, under the judicious management of Mr. Philip Davis, have alone carried on their operations throughout the entire year. The working of this mine, poor as it may be, is productive of enormous advantage to this District, where reefs abound in every direction that, if efficiently worked, will yield from five to ten pennyweights of gold per ton, and those who desire to do so can learn at the Dayspring workings how to make such reefs pay for the cost of extracting the gold, that is, pay wages and other expenses and leave a small margin for profit. The Dayspring Company has crushed, between the 1st of January and the 31st of December, 1874, 5,674 tons 13 cwt. of quartz raised from a mine 250 in depth and taken from a lode 2 feet 6 inches in thickness, invested by hard blue rock, and requiring to be removed by means of blasting powder; this stone produced 3,158 ounces 3 dwts. and 6 grains of gold, worth only £3 7s. 6d. per ounce; the average yield was 11 dwts. 12 grains per ton. The hardness of the investing rock and the narrowness of the lode necessitates an unusually large expenditure in working this mine. The entire cost of raising and crushing the stone and extracting the gold is £1 11s. 2d. per ton, the value when obtained is £1 18s. 3d., leaving a net profit of 7s. 6d. per ton.

As the Dayspring is a fair representative of the reefs in the District, it is here made manifest that these reefs and veins can be profitably worked by co-operative companies of working miners, provided that they are sober and industrious, and manage their affairs with economy.

3,293 ounces 3 dwts. 6 grains of retorted gold, the produce of the Dayspring and Ben Nevis Reefs, have been remitted to Sydney by Escort during the past year.

The produce of the Billabong Gold Field was—

	ozs.	dwts.	grains.	
Transmitted by Police Escort between the 1st of January and 31st December, 1874 ..	45,495	7	14	
Held by the Commercial Banking Company, Parkes, 31st December	2,402	2	14	
Received by do. do. up to 12th January, 1875 ...	625	18	20	produce of 1874.
Held by Australian Joint Stock Bank, Parkes, up to 31st December	615	13	0	
Received by do. do., up to 12th January, 1875 ...	522	4	4	produce of 1874.
In private hands and transmitted by private conveyance sundry times... ...	1,500	0	0	
Total produce	51,161	6	4	

The population on the Gold Field during the past year has fluctuated between 8,000 and 10,000 persons of all ages. These reside chiefly at Parkes and five small towns that have sprung up near the most important leads.

There are six steam quartz-crushing plants, containing all the most modern appliances for saving gold; in the aggregate 122 horse power drives 72 stampers. But one of these machines has been employed in the extraction of gold, one is used as a saw-mill, and four are idle. There are also 47 puddling machines, 115 whips and four whims.

Sixteen reservoirs of a temporary construction are held in connection with machinery, and seven others for general purposes; nearly all of these are shallow. 18 water rights are held by machine owners and 16 catch races convey storm water to as many small reservoirs for mining purposes.

The Forbes Gold Field has been imperfectly prospected and inefficiently worked, and I am of opinion that if the lead recently reported at the Bald Hill proves to be valuable the whole of the abandoned leads round Forbes, and extending from the banks of the Lachlan River, both on the northern and southern bank, for many miles will be re-occupied and re-worked with profit.

The unusual richness of the leads in the immediate vicinity of Forbes caused the miners of that day to view with contempt or indifference any auriferous deposits that were merely remunerative, consequently many leads were abandoned then that will be eventually thoroughly worked.

There is one quartz-crushing plant upon this field which has not been in operation during the year, there are also four puddling machines—they have been unemployed.

The Mining Registrar at Cargo reports:—

The most prominent feature of the Cargo Gold Field is the Ironclad Range, which is of considerable height, about 600 feet, and runs nearly south south-east and north north-west. This range seems to have been the feeder of the Gold Field, as on it all the principal reefs are situated, and in nearly all the ravines leading therefrom gold in payable quantities has been found.

The gullies on the eastern side which have been worked with good results are as follows:—First, Long Gully, which was the first place rushed on this field, and which supported a mining population of from 100 to 500 miners for nearly two years. The next on the same side of the range is Township Gully, which was opened about the same time as Long Gully; it was also highly remunerative, one of the leads running down the centre of Cargo-street. The next was Graveyard Gully, which paid well for working.

On the western side of the range are:—First, Scrape Rock and Tin-dish Gullies. These gullies were very shallow, and were soon worked over and produced a considerable quantity of gold. The next is Copper Gully, on which a good deal of work was done, but the amount of gold was not so great as in the others.

All these leads have been abandoned for a time, but there is no doubt several of them will be reworked and with good results. The reason of their abandonment was gold having been struck in Gum Flat, another gully on the same side of the range. When this gully was opened there was but a small mining population on the field, and the whole of the miners betook themselves to the new rush, where nearly all of them have remained ever since; the newcomers finding they could not get in on Gum Flat, did not care to set in to old ground of which they knew nothing, consequently left.

Gum Flat was opened upwards of three years ago; the gold was first struck in a gully leading thereto, at a depth of 30 feet; from thence it was traced down the flat on which there were several rich claims; Livemore and party had the best, some of their washings going as high as 4 ozs. to the load, the depth varied from 30 to 50 feet, it was then for a time considered almost worked out, when a party tried to sink through the false bottom on which the gold had been got, at a depth of about 90 feet, came upon a vein of wash almost perpendicular or dipping slightly like a reef, and on this vein (which has been traced through nine claims) has the principal work been done for alluvial during the past year.

Hicks and party's claim is supposed to be the best at present. This claim is somewhat different from the others; besides having the vein or lode, the same as the others, they have a flat bottom, at about 100 feet, over their entire claim, which yields well; and on this flat bottom they have been at work during the past year. Their best washing was 1 oz. to the load.

Another very important claim is O'Donald and party's. This claim, although not so rich in gold as Hicks's, pays almost as well, the vein being wide and easily worked.

Mackey and party's claim is the oldest claim on the flat. It has been worked by the present shareholders for upwards of three years, during which time it has paid from £3 to £7 per week per man. They worked it all over at about 45 feet deep, and now have commenced on the vein, which promises to pay them equally well.

Odgen and party's claim has been more difficult to work than the others, the ground being harder and the vein narrower, and pitching about from side to side, making it difficult to follow. They have, however, persevered, and followed it down from 90 to 245 feet. This claim has paid well; and at their present depth the prospects are much better than they have been. This is the greatest depth to which the vein has been traced in any of the claims.

Three other claims, viz., Holden's, M'Kay's, and Hogan's, have paid wages, and their prospects are improving.

Further down the flat is Rickey and party's. They got the vein at 40 feet, and are following it down. This is really a good claim, the vein being wide, easily worked, and above the average in gold.

Next to them is Groat and party. This claim is not so rich in gold as Rickey's, but is very easily worked, and a great quantity of stuff is got, which averages about 7 dwts. to the load. The vein has not been traced further than this claim.

It is impossible to say the exact quantity of stuff washed from these claims during the year; but, as near as can be ascertained, there has been 6,000 loads put through the machines, for an aggregate result of 2,800 ozs.

There is considerable speculation as to the future of Gum Flat, the formation being different from anything the diggers have experienced; it is the general belief, however, that the vein or lead extends a great deal further both ways than has yet been proved, the difficulty in finding it being very great, although the ground is easily sunk upon and worked, yet the vein being nearly perpendicular, with the same formation on each side of it, there is nothing to guide the prospector; he may sink within a few feet of it, and miss it; and again, in many places the vein is so narrow and poor that he may drive through it and not be aware of it. Some are very sanguine that this vein will lead them down to the main granite formation, where they expect to get something good. They seem to be assisted to this belief by the opinion expressed by the Rev. Mr. Clarke, that "on the western slope of the Canoblas there are heavy deposits of gold, but at great depth." There was a prospecting shaft put down on this Flat, upwards of 300 feet, without finding the granite bottom; it was abandoned for want of means.

There are a considerable number of known payable reefs on Cargo, nearly all of which are held under application to lease. They are as follows:—

Ironclad Reef,	Homeward Bound Reef,
Adelaide Reef,	Wreath of Roses Reef,
Pride of Cargo Reef,	Lucknow Reef,
Victim Reef,	Alpine Reef,
Dalcooth Reef,	Galatea,
Rise and Shine Reef,	Mobbs's Reef,
New Chum Hill Reef,	Prince Alfred Reef.

The owners of nearly all of these leaseholds are working on the alluvial on Gum Flat, and leave these tracts unworked for various reasons. First, they have remunerative employment where they are, and are not compelled to work these holdings until the leases are issued. Another reason is, during the first six months of the year there was only one crushing plant on the field, viz.:—the Rose of Denmark, the property of the Ironclad Company, who were nearly all the time crushing from their own ground, and when they did crush for the public they charged such a high rate that it cost 25s. per ton, viz., for carting and crushing, at which rate no poor reef would pay. There is now, however, another plant on the field, that of Wickins & Co., which was completed five months ago, but during three months of that they were unable to work from want of water; there is, however, a considerable quantity of quartz lying at several of the leaseholds, which will keep them constantly employed for a length of time. As soon as the leases are issued there is no doubt there will be ample employment for two more plants on the field.

The vein in the Ironclad Company's leasehold runs nearly N.N.E. and S.S.W.; it has been worked to a depth of 178 feet, and 200 feet along the line of the vein; it is nearly perpendicular; whatever dip it has is to the westward, it is in many places 12 feet wide, but will average 6 feet. There has been employed on this property during the past year from thirty-five to forty-five men; they have an excellent whim on the ground; the crushing plant has fifteen stamps, with a 25 horse power engine; there is a buddle attached which effectually separates the pyrites, of which they have a very great quantity; they have sent some of these to England to be treated; but are now about to erect a furnace and will treat them on the ground. The company will not, however, give the slightest clue as to what gold, copper, or silver they get; they have expressed themselves well satisfied, and everything about the property indicates that it is paying very well. They have raised and crushed during the year about 4,000 tons, and it is the general opinion that it has averaged an ounce to the ton.

The following is the result of the different crushings from this property previous to its purchase by the present company, viz.:—

	oz.	dwt.	
1st crushing yielded	5	10	per ton.
2nd "	5	15	"
3rd "	6	12	"
4th " (being 2 tons at the Mint)...	13	0	

About 30 tons were sent to England, and yielded 16 ozs. gold, 9 ozs. silver, and 35 per cent. of copper; there was also a large quantity of second class stone crushed, which yielded about 1 oz. 5 dwts. per ton.

The total quantity of quartz crushed on the field during the year has been, as near as can be ascertained, 6,000 tons, for an aggregate yield of 5,000 ozs.; this, together with the alluvial gold, viz., 2,800 ozs., makes the total yield of Cargo Gold Field for the year 1874 7,800 ozs.

There are two crushing plants on Cargo, with an aggregate of 37 horse power and 21 head of stamps.

There are nine dams on the Gold Field—three in Copper Gully lately erected, supplying Wickin and party's crushing plant—one dam and a large reservoir supplying the Ironclad Company's plant—four dams in Cargo Creek, supplying three puddling machines which are constantly employed on the alluvial from Gum Flat; there is also a dam in Township Gully for domestic use.

There are six leases of copper tracts on the Gold Field, none of which have been as yet worked.

There are on nearly all the ranges round Cargo very strong indications of both gold and copper, none of which have been proved; in fact there has been very little prospecting done in the district; there are at present, however, two parties out prospecting, one towards Toogong and another towards Cadia.

The number of miners on Cargo during the year has been 150, and the entire population 400.

On Canowindra there are a number of payable reefs both held under application to lease and as claims; they are as follows, viz.:—

Blue Jacket.
Hayse's.
Queen of the Ranges.
Homeward Bound.

There is one crushing plant of eight horse power and five stamps.

There has been about 2,000 tons of quartz raised and crushed during the year, with an average yield of 17 dwts. to the ton, giving a total of 1,700 ozs. as the product of the place during the year.

There are no alluvial workings on Canowindra.

There is a population of about forty-five miners employed on the reefs.

On Toogong and Boney's Rocks there are several payable reefs held principally under application to lease; the reefs are as follows, viz., American, London, Maher's, Bevin's, Little Leader, and Flat Reefs.

There is one crushing plant, lately erected, of ten horse power, and eight head of stamps.

They have crushed during the short time the machinery has been erected about 150 tons, for an average of 12 dwts. to the ton.

Mr. Warden Browne reports:—

1. Alluvial mining is still carried on with considerable energy and success in the Mudgee Mining District, but, as heretofore, by far the larger proportion of the gold sent down by fortnightly Escort is procured from the auriferous drift of Gulgong and Home Rule proper, or of later discoveries within a circuit of twenty miles. The mining population is more diffused, and the total amount of gold raised less than at the commencement of these Gold Fields, but the returns of 1874, viz., 72,488 ozs. 4 dwts. 10 grs., afford proof that the exceptionally rich deposits referred to are still far from total exhaustion.

2. The machinery employed for purposes connected with gold-mining, though inadequate to the full development of the deeper alluvial leads, yet will be found to represent no inconsiderable amount of capital. It comprises (20) twenty steam-engines, of an aggregate amount of 264 horse power, with (104) one hundred and four puddling machines, (42) forty-two whims, and (135) one hundred and thirty-five "whips,"—all worked by horses. This list is exclusive of water-wheels, sluice-boxes, and hydraulic hoses. It may be appropriate to mention here that costly machinery is at present employed upon the lower portion of the Moonlight, the Star, and the Black Leads, as also at the prospecting claim of the Buchanan. At the Great Extended Black Lead Company's lease—where powerful steam machinery is employed for pumping—I am informed that very encouraging returns have been lately received. I share, with persons of experience and geological eminence, a fixed belief in the future discovery of rich deposits underlying the basaltic formation at a greater depth than has been hitherto reached.

3. Seventy-six (76) leases have been applied for during the year 1874 at Gulgong, in occupation of 407 acres of alluvial, and (64) sixty-four acres of quartz reefs or veins. I am unable to point to any instance of proved richness in quartz-mining, though it should be stated that capital, at Gulgong, has never yet been applied, save very sparingly, to the efficient testing of promising reefs which are known to exist. An exception to this rule may yet be found in a line of reefs, discovered recently, at the abandoned alluvial diggings of Three-mile. One of the working parties there had a crushing so successful of late that a continuation of the average yield will at once, and favourably, settle the question of the existence of rich quartz reefs in the vicinity of Gulgong. In the neighbourhood of Hargraves and Windeyer, as at Apple-tree Flat, near Mudgee, there is a fair amount of labour employed and capital expended in quartz-mining, though, as in the vicinity of Gulgong, by far the larger part of the population is concerned with the alluvial.

4. With reference to minerals other than gold, I am not aware of any prominent discovery excepting that of the Belara Copper Mine, distant about twenty miles from Gulgong in the direction of Wellington. The indications are encouraging, and an assay of the grey ore, or of the red and black oxides, has shown a good percentage, but no continuous work, leading to certainty of opinion, has yet taken place.

5. Considerable migrations have taken place to the Palmer and to other Gold Fields, but a noticeably large proportion of miners have in every case returned, demonstrating the general confidence felt in the permanent character of the Gold Fields of this district.

6. The protracted dry weather has been highly unfavourable both to prospecting parties and to the general yield of gold, as large accumulations of wash-dirt have been recently awaiting sufficient water for the puddling process; but rain having lately fallen in large quantities, I have every confidence in a succession of discoveries of payable gold, and in a marked increase of the general yield.

Name of Lead.	Depth of Wash.	Width of Wash.	Average yield per load.
GULGONG.			
BLACK LEAD.			
150 ft. sinking—alluvial....	6 in. to 12 in.	70 ft. to 300 ft.	4 to 6 dwts.
CALEDONIAN LEAD.			
85 to 100 ft. sinking	4 in. to 6 in. ...	100 ft. to 600 ft.	5 to 8 dwts.
HOME RULE.			
CHRISTMAS LEAD.			
80 ft. sinking (worked out)—coarse gold.	2 ft. 6 in. ...... (quartz wash.)	50 ft.	15 dwts.
HOME RULE LEAD.			
92 ft. to 170 ft. sinking—fine gold.	½ in. to 2 ft. ... (fine wash.)	150 ft.	10 dwts.
SHALLOW RUSH.			
9 ft. to 125 ft. sinking—coarse gold.	2 ft. .... (fine wash.)	100 ft.	1 oz.
LOWE'S PADDOCK.			
80 ft. sinking – mullocky wash with coarse gold (worked out).	3 ft.	400 ft.	½ oz.
CANADIAN AND NIL DESPERANDUM LEADS.			
Two layers of wash—70 ft. to 158 ft. sinking.	3 ft. to 15 ft. (mullocky wash, coarse gold.)	10 ft. to 40 ft.	5 dwts.

At the Cement Claim at Old Tallaway, there are 10 to 25 feet of cement or conglomerate, yielding 3 dwts. to the ton.

Mr. Warden Buchanan reports:—

In the immediate locality of Armidale two of the gold-fields, which it was believed would prove remunerative if machinery was only at hand for crushing, have now been almost abandoned, and this after quartz-crushing machinery had been supplied by private enterprise. As far as I can learn, the reefs in the localities I refer to, namely, Glen Morrison and Cameron's Creek, might be payable enough if worked upon a large scale and with increased crushing power, but unfortunately the results, as obtained from but moderate appliances, have been so unsatisfactory that, upon the one hand, capitalists have become dispirited, and, on the other, the miners have either abandoned the ground or registered it. At present I see no immediate prospect of any reaction in the working of the reefs, and the alluvial fields are almost left to the enterprise of a few sluicing parties, who, in most cases, are working old ground.

The cause of the depression is not the actual poverty of the gold reefs, but the want of some more economical method of working them.

The tin fields, especially at Vegetable Creek and Cope's Creek, continue to yield steady returns, and the industry after much fluctuation may be said to be thoroughly established. Townships at both of these places are in the most flourishing condition, and there seems every prospect of many years elapsing before the ground actually in process of working will be exhausted.

I estimate the population on the tin fields at 2,000 men, and of this number the majority are either tributers or men actually employed by private companies. The existence of valuable lodes was satisfactorily proved some years ago, but as yet nothing has been done towards their development. The general opinion is that a far more economical method of extracting the ore must be discovered than at present known before the lodes can be worked to advantage, and that appliances for smelting on the ground must be introduced. The attention, however, of English capitalists has been drawn to the tin fields of New South Wales, and several blocks of land have been inspected by accredited English agents. Pending their official reports, which are stated to be satisfactory, being in the hands of their constituents, nothing can be definitely arranged. I contemplate, however, increased activity in the working of the tin mines during the ensuing year, and I may state that from my personal inspection and experience, I am convinced that a wide field is open to the labouring classes upon them.

Report of Mr. Mining-Surveyor Phillips, on the Nundle and Denison localities, the Upper Peel, and Hunter Rivers, New South Wales.

Geology and lithology.

The practical and scientific miner, however protracted and extensive his experience, may come here with edification in things strange. Over a wide range in Victoria "pipeclay," "cement," and "sandstone," with common quartz veins, exhaust the miner's vocabulary, or we may add the volcanic "bluestone" of recent occurrence. The name of either conveys to his mind a definite idea of a stone or mass of uninteresting and usual aspect, but in writing of the lithology of some parts of this side of New South Wales, and mountain sources of the river Peel especially, I feel at a loss for adequate expression, having omitted no opportunity of inspecting museums in furtherance of my vocation, and I can say that a day's inspection of the drifted matter of the river Peel bed presents more in variety than I can call to mind through life—porphyries, granites, slates, trachites, dolomites, silicates, calcites, steatites, serpentines, quartzites, jaspers, common quartz being of rarest occurrence—each of these in great variety, spotted, streaked, mottled, and of highly coloured aspects; some harder than steel, and tough withal, so that a hammer will not break them without extreme exertion; others exfoliating spheroidally, others soft and crumbling on exposure to air. Mottled soaps of different colours very strictly resemble the serpentine rock, and there is a veined serpentine of fiery beauty, not steatitic but opaline.

Foley's Folly Basin is a highly interesting glen for the lithologist.

There is a regular granite somewhere about the head of the Peel, but I did not come on it. The hill cone of columnar basalt hereabout is the most surprising specimen I have seen, as suggesting queries and inciting contemplation as to the how and wherefore—a high peak composed of columns piled like bricks one on another, without any broken or earthy matter between.

Gold habitat.

When it is observed that, in the absence of the ordinary quartz of auriferous rocks, this locality has been yielding gold for some sixteen or twenty years, there is an important inference that gold has no specific habitat. It is, moreover, seen here that the auriferous veins are composed as much of calcareous as silicious spar.

Payable gold.

The cements of this district clothing the ridge peaks have little or no quartz, but they promise payable gold for years to come.

Other ores.

I have found chromate and other iron ore in the presence of lime and fuel, maganese antimony, tungstate of iron, but not a stain of copper that I have met with.

Hypothetical.

Although the cements here are not quartzose, they are as I think of the same geological age as those so widely seen in the Gold Fields of Victoria, and are, as I believe, the earliest work of the sea, and not what they are called in the authorized Victorian geology. I have always been satisfied that they are coeval with and related to the "*Old Red.*" Those who question their great age may come here and inspect road cuttings, showing well defined trap dykes like walls up through the boulder-beds.

Anomalies.

There are reefs and veins younger than animals that once lived. A promising silver lode, with gold in fossiliferous rock, through Dr. Creed's and other adjacent land, in other parts the inferior precious stones taking the place of quartz with gold, such as agate, carnelian, chalcedony, coarse jasper, coarse opal, &c., proving, as I said before, that gold has no particular companionship or dwelling-place.

There is lime-rock with silicious lodes near Croney, quartz-stones with encrinite stems and other nuclei, either floral or faunal, removed.

There are pelagic beds—"cement" of the gold-miner—broken through by volcanic eruption. There is, remote from this, at Currajong, a sandstone or elvan dyke, many yards in width, between clay-slate walls, traversed obliquely by auriferous quartz veins, the latter being bounded or cut off in both directions by the slate-rock. At the same place there is a counter dyke in the larger one, and amongst the quartz veins composed of bluestone, much resembling the so-called volcanic stone, the last-named dyke, enclosing numerous well-rounded smooth boulders, much of the same sort of stone as that enclosing them; in other words, it is a *pudding-stone dyke traversing an elvan-stone dyke.* Although this is some 350 miles from the subject of this report, I have inserted it as a fact explanatory of the confused order of the rocks elsewhere seen.

Coming back to the Hunter and Peel—and in allusion to the pudding-stone or "cement" lying upon slate on edge—I have to notice the effect of a keen but unknown agency by which the fissures or cleavage of the slate underneath are produced up through the conglomerate; such cutting or penetrating action bisecting everything in its way, boulders hard and soft, like a sharp knife would cut through a plum-cake. Sections of these cuttings being now seen in the side gullies about the Denison and Moonam Brook—the cement rock showing smooth plane sides, with the hard pebbles cut through as with a saw. Further, and exceedingly noteworthy in a scientific point of view, the pudding-stone is sometimes sliced and re-cemented after a sliding motion as it were of slice on slice has had place; so that the several portions of boulders do not unite in a fitting manner, but with odds on each side, as elucidated by a specimen shown me by Dr. Creed, of Scone, and others shown me in England by Messrs. Fox, of Falmouth.

Hypothetical.

This metamorphism or cleavage of the "Old Red" or other strata on the Silurian, Permian or clay slates, may have been induced by accidental planetary heat; thence thermo-electricity, which may constantly have play in the earth's lithogenic arrangements.

Further, as it regards the ancient and recent auriferous valleys, and in order to explain their causes and non-conforming positions, we suggest that the undermost rocks of this part of the earth at first suddenly uplifted were eroded into gullies by the violence of the retiring ocean, and before there was time for floral life or for its remains to be collected in the gullies the continent again dropped partially under the sea, and so remained until drifted woody matter, ordure of fish, mud, and sand lying on the bottom of the sea or lake, became buried in lava, which by its own heat and help of hydraulic pressure, dissolved silex for the petrifaction of wood and coprolites now so extensively found underlying the hill-caps of lava. We have only then to suggest a final precipitation of the ocean in a westerly direction over this country to produce the present gullies with their gold, cut through everything, denude the eastern sides of mountains to the bare crag, clothe the western with diluvium, and thus carve out a new set of gullies with their golden drifts.

Mining enterprise.

In coming to the mining enterprise of this district, it may be said that it is not exempt from that moral blight by which mining and other human enterprises will continue to suffer as heretofore. There is a faunal, a floral, and mercantile blight by which man is threatened and punished; but he must nevertheless plant his crop, rear his live stock, and initiate new enterprises in the chances there are of escape.

Whatever the villainy by which mining is brought to a low ebb in this country, there is the consolation that the mines and the hills are intact, and remain for the discreet worker, the *bonâ fide* and patient capitalist; and there are inducements fair for the future. But that may have to be deferred until the extensive cultivation of the soil, manufactures, and increased population concur to assist mining. Labour being a great element in mining, a little reduced will greatly enhance the balance of profit; 10 per centum off labour might put thirty or forty on to profits. For instance—

Costs, including labour, £1,200; other costs, £300 ...	£1,500
Proceeds	1,600
Profit	£100

If said labour be reduced 12 per cent., it becomes £1,056; add £300	£1,356
Proceeds	1,600
Profit	£244

144 per cent greater.

Mining facilities.

The mining facilities in this part are very rarely exceeded; a systematic tunnelling commanding mine over head for many years, and gravitation being thereby brought into play.

Mining prospects.

If all the reefs within two miles of the Denison machinery belonged to one company, which would not be too much, there would be a fair chance of success among them all.

The Omadale Reef workings, properly initiated with the machinery there set up, are especially inviting.

The Moonlight, Starlight, Daylight, The Duke of Edinborough, and other reefs in proximity, along and near the Peel, being commanded by tunnels, might well engage the British capitalist.

If the leases on Foley's Reef, at Bowling Alley Point, the flat by Lindsay's, and the Kanacka's Reef, could be thrown into one concern, or even adding the Marquis of Lorne and surrounding grounds for a British mining company of untiring capital, there would be good and lasting mining—as we look at it.

If the notorious cement claim at Mount Misery, and ground of the same character within a mile or two of it, could be worked by an efficient company, placing their machinery and tunnels at the bottom of the hill and using water-power in lieu of steam, good might be done.

PLAN OF M^c GUIGGAN'S LEAD.

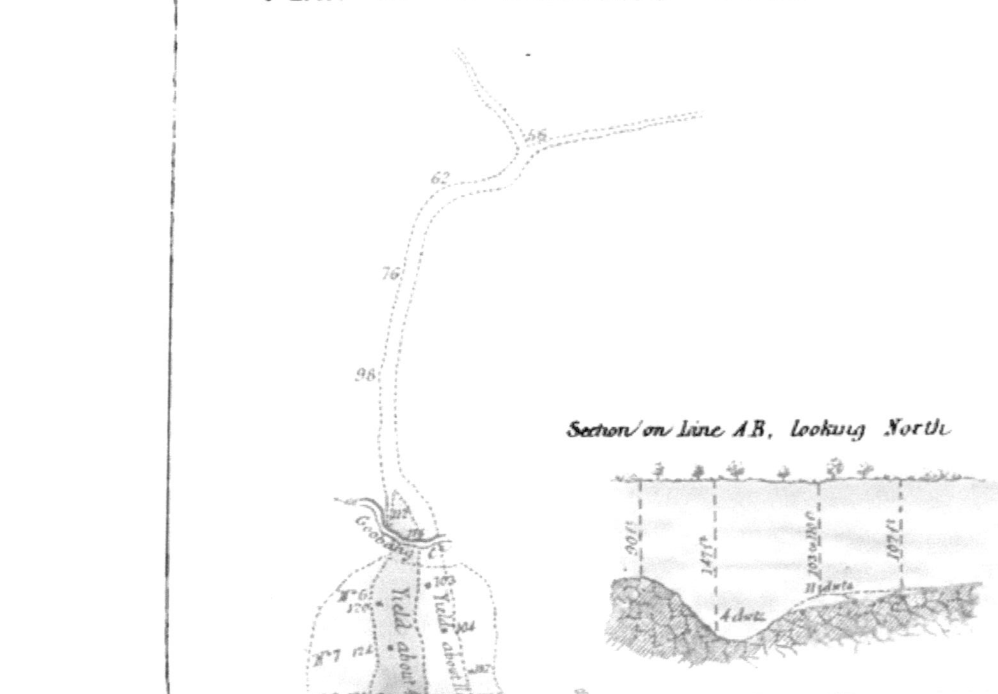

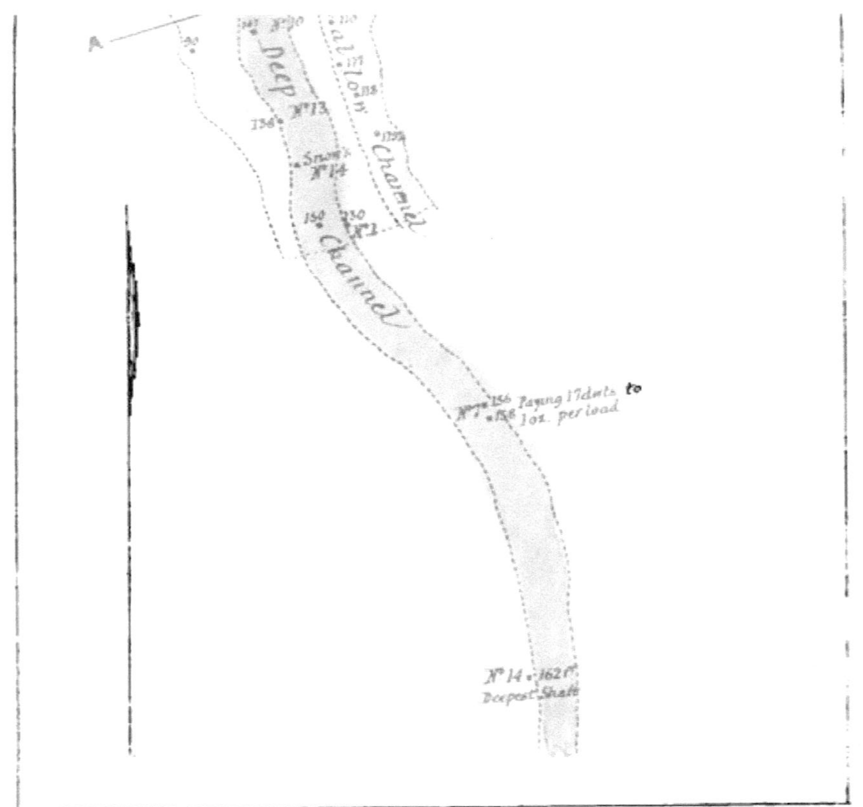

A
Deep
No 13
Snow's
No 14
Channel
Channel
Paying 17dwts to
1oz. per load
No 14 162ft
Deepest Shaft

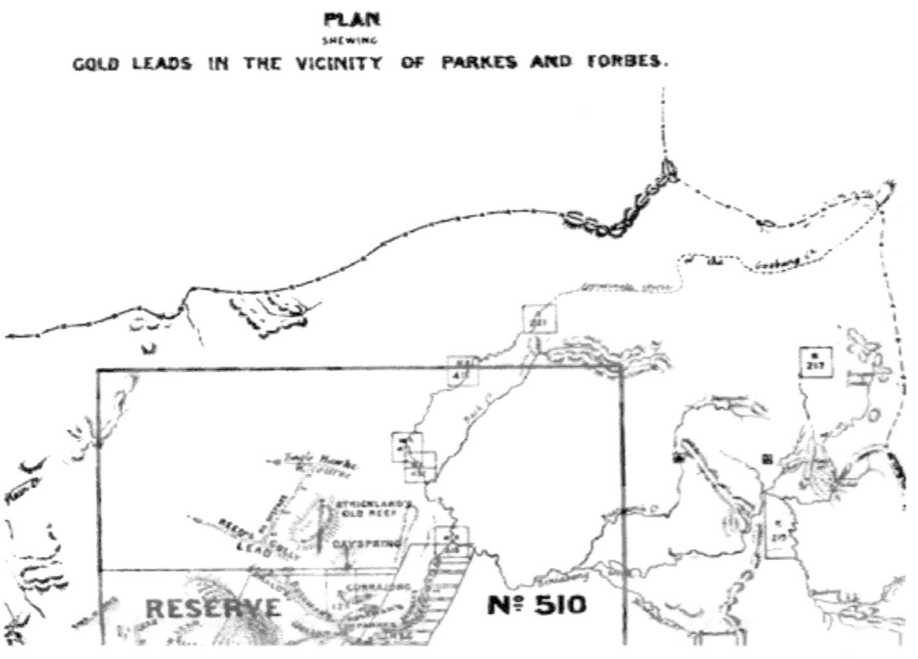
PLAN
SHEWING
GOLD LEADS IN THE VICINITY OF PARKES AND FORBES.
STRICKLAND'S OLD REEF
DAYSPRING
RESERVE
Nº 510

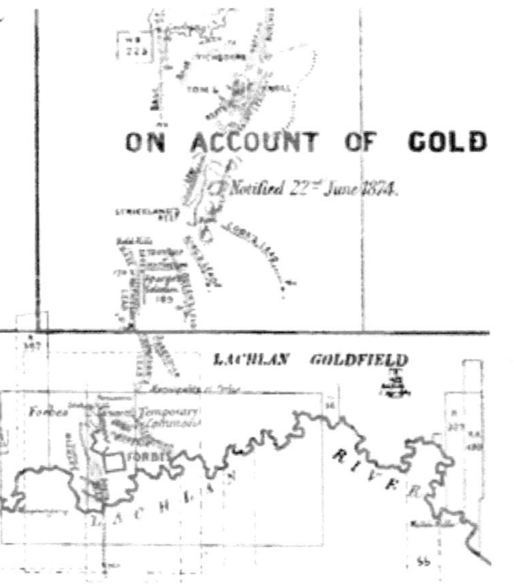
ON ACCOUNT OF GOLD
Notified 22nd June 1874.
LACHLAN GOLDFIELD
Temporary Commons
FORBES
Forbes
LACHLAN RIVER

Then, and lastly, as far as my inspections have extended, there is the great gold and silver course of veins, through and beyond Dr. Creed's land, as fairly inviting a trial on the part of the practical silver miner as anything I can think of in the Colony.

The reason of my alluding especially to the British mining capitalist is, that he knows that time, as well as money, is essential to the development of mines, and that he is patient to await a right pioneering of ground, and that although the first spot tried fails he must try again; that he has to go through a scope of ground, taking the poor with the good; whereas, the colonial capitalist, who has numerous other means of speedy investment and sure returns in his business, learnt and understood and better paying, cannot wait.

Gratuitous power.

There are many places in this district where 300 feet or more of waterfall can be had; and if the water-piston were brought to bear in aid of its mining by a competent mechanic it would open the eyes of interested parties to a tremendous power lost sight of, and to a corresponding amount of money that might have been saved.

The balance bucket principle has been recently introduced here by Mr. Stanning, the Manager of the Foley's Folly mine, for a specific purpose, while the advantage of the piston over everything else is, that it is adapted to all work, cheap, and safe from explosion or other accident.

COAL.

The following statement of the Coal raised since the year 1829 shows the progress made in this branch of mining. Large tracts of coal land have been taken up with a view to open new collieries, and there is every reason to believe that ere long the quantity of coal raised in this Colony will far exceed the quantity being raised at present. The extent and value of our coal beds are so great that the supply may be said to be practically inexhaustible, and the demand, both for home consumption and for export, is likely to increase to such an extent as to be limited only by the raising power of our collieries and the facilities for bringing the coal to market.

Coal.			Coal.		
Year.	Quantity.	Value.	Year.	Quantity.	Value.
	Tons.	£		Tons.	£
1829	780	394	1852	67,404	36,885
1830	4,000	1,800	1853	96,809	78,059
1831	5,000	2,000	1854	116,642	119,380
1832	6,000	2,100	1855	137,076	89,082
1833	328	124	1856	189,960	117,906
1834	8,490	3,750	1857	210,434	148,158
1835	12,392	5,483	1858	216,397	162,162
1836	12,616	5,747	1859	308,213	204,371
1837	16,083	7,828	1860	368,862	226,493
1838	17,220	8,399	1861	342,067	218,820
1839	21,283	10,441	1862	476,522	305,234
1840	30,256	16,498	1863	433,889	236,230
1841	31,841	20,905	1864	549,012	270,171
1842	39,900	23,940	1865	585,525	274,308
1843	25,862	16,222	1866	774,238	324,049
1844	23,118	12,363	1867	770,012	342,655
1845	22,324	8,769	1868	954,231	417,809
1846	38,965	13,714	1869	919,774	346,146
1847	40,732	13,750	1870	868,564	316,836
1848	45,447	14,275	1871	898,784	316,340
1849	48,516	14,647	1872	1,012,426	396,198
1850	71,216	23,375	1873	1,192,862	665,747
1851	67,610	25,546	1874	1,304,567	790,224
			Total ...	12,387,279	£6,655,328

STATEMENT showing the total quantities and value of Coal raised from the undermentioned Mines.

Name of Colliery.	Where situated.	Coal raised prior to 1874. Quantity.	Coal raised prior to 1874. Value.			Coal raised during the year 1874. Quantity.	Coal raised during the year 1874. Value.			Total Quantity.	Total Value.		
		Tons.	£	s.	d.	Tons.	£	s.	d.	Tons.	£	s.	d.
Glen Rock	Newcastle	2,977	1,562	10	0	1,400	735	0	0	4,377	2,297	10	0
Victoria Tunnel	Newcastle	1,487	780	13	0	2,148	1,127	14	0	3,635	1,908	7	0
Dark Creek, Jesmond	Newcastle	700	280	0	0	150	75	0	0	850	355	0	0
Waratah	Newcastle	1,202,933	421,026	0	0	181,268	107,032	0	0	1,384,201	528,058	0	0
Newcastle, Wallsend	Newcastle	1,900,000	950,000	0	0	240,000	163,000	0	0	2,140,000	1,113,000	0	0
Co-operative	Newcastle	112,609	71,766	0	0	149,699	94,982	0	0	262,308	166,748	0	0
Greta	Maitland	300	210	0	0	29,030	20,321	0	0	29,330	20,531	0	0
Osborne, Wallsend	Wollongong	38,422	16,329	0	0	37,796	16,063	6	0	76,218	32,392	6	0
Mount Pleasant	Wollongong	35,749¾	15,193	13	9	38,985	16,568	12	6	71,734¼	31,762	6	3
Mount Kembla	Wollongong	8,000	4,000	0	0	1,000	500	0	0	9,000	4,500	0	0
Lithgow Valley	Hartley	2,865	859	10	0	18,000	5,400	0	0	20,865	6,259	10	0
Eskbank	Hartley	17,388	6,155	16	0	8,600	3,010	0	0	25,988	9,165	16	0
New Wallsend	Brisbane Water	400	380	0	0	20,000	12,500	0	0	20,400	12,880	0	0
Rix's Creek	Singleton	130	91	0	0	180	126	0	0	310	217	0	0
Mount Wingen	Murrurundi	361	171	0	0	None	raised.			361	171	0	0
Rockroof	Berrima	350	210	0	0	350	210	0	0	700	420	0	0
		3,321,671¾	1,189,015	2	9	728,606	441,650	12	6	4,053,277¾	1,930,665	15	3

KEROSENE SHALE.

The produce of the Colony of New South Wales.

Year.	Quantity.	Value.
	Tons.	£ s. d.
1865	570	2,350 0 0
1866	2,770	8,154 0 0
1867	4,079	15,249 0 0
1868	16,952	48,816 0 0
1869	7,500	18,750 0 0
1870	8,580	27,370 0 0
1871	14,700	34,050 0 0
1872	11,040	28,700 0 0
1873	17,850	50,475 0 0
1874	12,100	27,300 0 0
	96,141	£261,414 0 0

RETURN of Kerosene Shale Mines in the Colony, and the quantity obtained from the same, and its value, in the year 1874.

Name of Mine.	Where situated.	Quantity produced and value.	
		Quantity.	Value.
		Tons.	£ s. d.
Greta	Maitland	100	300 0 0
Mount Kembla	Wollongong ..	3,000	4,500 0 0
New South Wales Shale and Oil Company's	Hartley	9,000	22,500 0 0
		12,100	27,300 0 0

The Examiner of Coal Fields reports :—

The quantity and value of the coal raised in the different districts during the year 1874 is as follows :—

Newcastle District.

Bituminous coal used for steam, household, smelting, gas, blacksmith, and coking purposes :—

	Tons.	Value.		
Newcastle Wallsend Colliery... ...	240,000	£163,000	0	0
Australian Agricultural Co's... ...	195,494	120,963	0	0
Co-operative Colliery...	149,699	94,982	0	0
Waratah Colliery	181,268	107,032	0	0
New Lambton Colliery	133,805	84,815	15	3
Lambton Colliery	127,768	80,488	16	0
Duckenfield Colliery	3,821	2,594	4	0
Victoria Tunnel	2,148	1,127	14	0
Glen Rock Colliery	1,400	735	0	0
Total quantity and value in 1874...	1,035,403	655,737	17	0
" " 1873...	1,014,223	581,801	13	0
Increase in 1874	21,180	£73,936	4	0

against 155,507-17 tons, valued at £240,827 13s. (increase), in 1873.

Four-mile Creek and Branxton, &c., in the Northern District—Splint and bituminous coals suitable for steam, household, gas, smelting, blacksmith, and coking purposes:—

	Tons.	Value.		
Pearse and Co., Four-mile Creek ...	11,088	£3,218	0	0
Ingance Colliery " ...	5,858	2,636	1	0
Sunderland " ...	1,200	360	0	0
Bloomfield " ...	757	106	18	0
Dark Creek " ...	150	75	0	0
Greta Coal and Shale Company ...	29,030	20,321	0	0
Anvil Creek Colliery Company ...	24,000	16,800	0	0
Rix's Creek, near Singleton... ...	180	126	0	0
Stony Creek, near Maitland	500	300	0	0
Total quantity and value in 1874...	72,763	43,942	19	0
" " 1873...	31,280	17,381	6	0
Increase in 1874	41,483	£26,561	13	0

against 8,022 tons, and £13,358 value (increase), in 1873. This is entirely owing to the increased vend at the Greta and Anvil Creek Collieries.

Greta Company, petroleum oil cannel coal, 100 tons, valued at £300, being the same quantity as that sold in 1873.

Catherine Hill Bay, near Lake Macquarie.

Splint and bituminous coal used for steam, household, smelting, and blacksmith purposes.

	Tons.	Value.		
New Wallsend Colliery Company ...	18,147	£11,795	11	0
Total quantity and value in 1873 ...	400	380	0	0
Increase in 1874 ...	17,747	£11,415	11	0

Southern or Illawarra District.

Semi-bituminous coal used for steam, household, smelting, and blacksmith's purposes.

	Tons.	Value stated unknown say—
Bulli Colliery	58,506	£29,253
Mount Pleasant Colliery	38,985	16,568
Osborne Wallsend Colliery	37,796	16,063
American Creek, used for oil-making ...	1,000	500
Total quantity and value in 1874	136,287	62,384
" " 1873	137,062	62,889
Decrease in 1874	775	£505

	Tons.	Value.
American Creek, petroleum oil shale, made into oil at the Works	3,000	£4,500
Total quantity and value in 1873	2,750	4,125
Increase in 1874	250	£375

against 10 tons, valued at £25 (increase), in 1873. No new mines have been opened out in this district during 1874.

Southern District.

	Tons.	Value.
Brereton's Coal Mine near Berrima	1,000	£1,250

This mine has been reopened during the year 1874, but I have not received the usual notice of this mine having commenced work again, and must inquire into it.

Western District.

Lithgow Valley, Hartley, and Mudgee Road. Splint coal used for household, steam, smelting, gas, blacksmith and coking purposes:—

	Tons.	Value.
Lithgow Valley Colliery...	18,000	£5,400 0 0
Thos. Brown, Esq., M.L.A., Esk Creek Colliery	8,600	3,010 0 0
Bowenfels Colliery Company	8,500	no value given say 2,975 0 0
Vale of Clydd Company	50	17 10 0
Bulkeley's Coal Mine at Blackman's Flat, Mudgee Road	50	20 0 0
Total quantity and value in 1874 ...	35,200	11,422 10 0
" " in 1873 ...	9,865	3,253 17 6
Increase in 1874	25,335	£8,168 12 6

against 4,641 tons, valued at £1,088 17s. 6d. (increase), in 1873

		Tons.	Value.
New South Wales Shale and Oil Company— Petroleum oil cannel coal, used for oil and sold for gas purposes in 1874	...	9,000	£22,500 0 0
Total quantity and value in 1873	...	15,000	46,250 0 0
Decrease in 1874	...	6,000	£23,750 0 0

The above return for 1874 was furnished me by the Mining Department.

Recapitulation showing the quantity extracted from the whole of the Mines.

There were twenty-eight collieries raising coal and three getting petroleum oil cannel coal and shale, and the aggregate production of coal from these collieries in 1874 was 1,298,100 tons, valued at £786,152 17s.

The aggregate production of petroleum oil cannel coal and petroleum oil shale in 1874 was 12,100 tons, valued at £27,300.

Return of the Number of Coal Mines and Quantity and Value of Coal raised, from the years 1864 to 1874 inclusive.

Year.	Number.	Quantity.	Value. £	s.	d.
1864	25	549,012½	270,171	11	0
1865	24	585,525½	274,303	13	9
1866	25	774,238	324,049	6	7
1867	26	770,012½	342,655	7	8
1868	28	954,230¼	417,809	6	1
1869	33	919,773¼	346,145	16	5
1870	32	868,564½	316,835	16	4
1871	27	898,784½	316,340	2	1
1872	26	1,012,426¾	396,197	19	10
1873	29	1,192,861¾	665,746	17	3
1874	28	1,298,400	786,152	17	0

From the foregoing returns we find that our coal trade is year by year increasing in a most satisfactory manner, and has never been in such prosperous condition as it is at the present time. Many new Companies have been formed, as well as very large areas of coal land taken up in various parts of the Colony, with the intention of working the coal from under it. If this rapid increased demand for our coal could have been foreseen a few years ago, and the shipping facilities at Newcastle had been greater than they now are, we should have had a much larger production and demand to report; and when the extra wharfs and cranes now in course of erection at the Newcastle Harbour are completed, there will be a very much larger foreign demand for our coal.

The agreement entered into by the associated masters and the officers and delegates of the Coal Miners' Association of the Hunter River District, by which the wages paid for hewing coal, and other work usually done by the miners, the hours of labour to be observed at the different collieries, and the mode of settling any disputes that may arise in reference thereto, are to be arranged, is working well, and there is no doubt about its having been the means of keeping the price of coal at 14s. per ton delivered into vessels in Newcastle harbour.

I annex a very interesting Return of the Newcastle Foreign and Intercolonial Trade, compiled and kindly given to me by Mr. Logan, the Newcastle Collector of Customs.

I am now preparing for the Philadelphia Exhibition plans showing the position of the different collieries in New South Wales, with the outcrop of the seams of coal shown thereon; also, sections to illustrate the thickness of the seams of coal, and the part worked in all our principal coal mines; as well as a longitudinal section of the lower coal measures near Strond, in which there are Sigillaria, Stigmaria, &c.

As soon as I have finished them I propose to make you a Supplementary Report for the past year, which will contain an account of the whole of our Coal Fields, and the new mines opened out in the year 1874.

NEW SOUTH WALES.—PORT OF NEWCASTLE.

Foreign and Intercolonial Trade.

Year.	Entered Inwards from Foreign and Intercolonial Ports.		Cleared Outwards for Foreign and Intercolonial Ports.		Total Value of Imports from Foreign and Intercolonial Ports.	Quantity and Value of Coal exported to Foreign and Intercolonial Ports.		Total Value of Exports (including Coal) to Foreign and Intercolonial Ports.	Total Amount of Revenue collected.
	No. of Vessels.	Tonnage.	No. of Vessels.	Tonnage.		Tons.	Value.		
					£ s. d.		£ s. d.	£ s. d.	£ s. d.
1864	664	196,961	795	266,528	59,656 2 0	299,150	144,748 0 6	248,316 9 6	16,555 18 9
1865	676	189,620	872	248,769	78,355 17 0	302,362	142,159 9 0	238,972 6 2	24,203 5 2
1866	799	246,346	992	308,575	53,219 3 5	411,746	184,132 16 9	216,177 17 5	29,959 15 8
1867	688	229,064	925	303,504	98,083 6 1	398,022	182,288 1 9	209,949 16 3	29,793 15 0
1868	871	296,517	1,100	372,718	84,486 15 5	480,069	226,440 0 0	283,783 5 0	31,175 9 5
1869	854	297,855	1,084	386,176	151,410 4 5	503,866	223,566 4 2	252,124 3 6	33,058 0 5
1870	765	283,091	1,046	383,242	154,816 5 8	511,545	223,077 7 0	241,435 16 8	32,145 5 7
1871	745	277,959	1,040	376,378	203,168 2 7	489,714	208,833 9 0	236,683 9 3	26,590 1 9
1872	876	342,514	1,092	427,845	268,141 12 11	565,994	243,911 18 0	282,834 9 10	41,196 9 8
1873	978	389,121	1,259	498,468	310,101 11 11	650,899	412,631 5 9	591,032 6 6	48,864 16 8
1874	1,156	510,291	1,269	543,693	343,297 19 11	* 723,844	496,448 15 0	697,048 7 7	59,387 7 11

* NOTE.—In addition to the above, 276,317 tons of coal were shipped coastwise in the year 1874.

TIN.

The only portion of our stanniferous districts from which any satisfactory report for the past year has yet been obtained is Vegetable Creek (*vide* Mr. Gower's report annexed). That large quantities of stream tin have been raised at Cope's Creek, Tingah, is well known, but whether the subjoined statement embraces all the tin ore that has been raised is more than doubtful. During the year 1874, the quantity and value of the tin ore won in the Maryland District is said to be:—Maryland, 2,182 tons; value, £107,000. Mole Tableland, 246 tons; value, £14,022. Total, 2,428 tons; value, £121,022. There is no record of the tin ore raised in the southern part of this Colony during the year 1874, except that the Pulletop Company, near Wagga Wagga, raised 1,200 lbs., value £40. The bulk of the tin from Maryland and some other fields in the north is sent into Queensland, and that raised in the south into Victoria. A return of the quantities taken across the Border has been kindly furnished by the Customs Department, and is included in the subjoined table.

The progress made in washing stream tin is most satisfactory, and there appears amongst the tin-miners a strong desire to avail themselves of every means of reducing the cost of washing. In many places the tin is washed in sluices, but in others the earth containing the tin has to be puddled. It is questionable whether the use of an iron puddling machine, having an opening in the bottom, through which the puddled dirt might be made to fall direct into the sluice, or into the feeder of the cylindrical sieve, without shovelling, would not be found economical. The great objection to the iron machine will probably be the original cost, but it is thought the saving of labour will more than compensate for the extra outlay.

Until a comparatively recent date the operations of our tin-miners were restricted to the working of the beds and banks of existing watercourses, and of other deposits near the surface; but of late, beds of ancient streams, 60 or 80 feet below the surface, containing rich deposits of tin ore, have been discovered, and are being worked. The mode of working these deposits does not in any respect differ materially from the working of the deeper auriferous leads. The discovery of these deeper deposits imparts to our tin fields an appearance of permanence that promises remunerative employment to a very large number of miners for many years to come, and will probably lead to the erection of a greater number of smelting works.

Mr. Wilkinson (the Geological Surveyor of this department) in his valuable Report upon the tin-bearing country in the district of Inverell, addressed to the Surveyor General, in 1872, directed attention to the older leads in the following words:— "The discovery of the older tertiary alluvial leads is doubt-

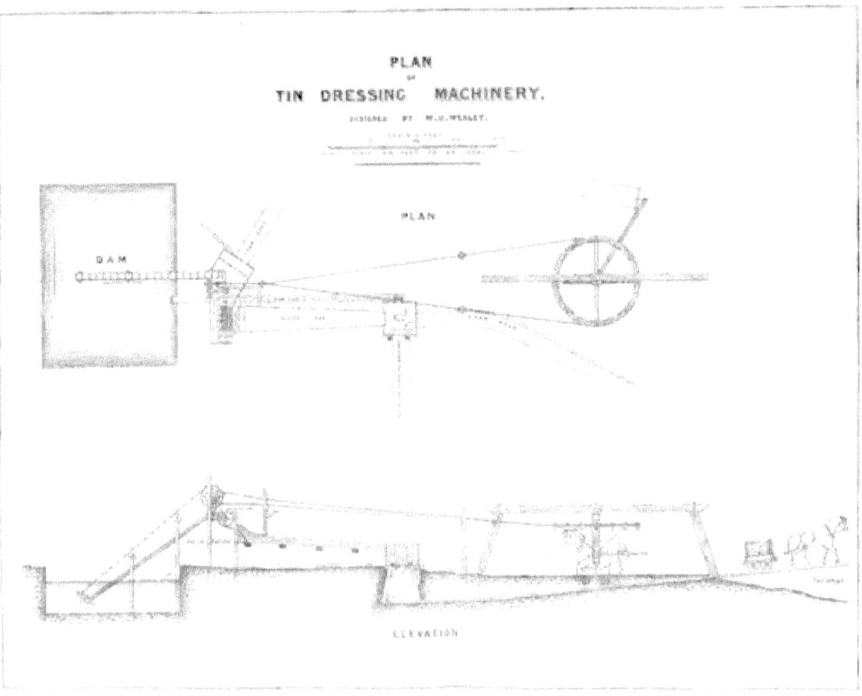
PLAN
TIN DRESSING MACHINERY.
PLAN
DAM
ELEVATION

less of much importance. When this old drift has been tested *in situ*, it has been found rich in coarse stream tin. If the actual channel of the old stream could be found, it might be found richly payable. But I would also draw the attention of the prospectors to where the present gullies have cut through the lead and redistributed its rich contents in their own channels a few hundred yards lower down. Where similar instances occur in Victoria, the beds of the existing gullies are sometimes found richer than that of the old lead they have cut through; they have as it were ground-sluiced the contents of the lead which once crossed the present valley, and have concentrated its rich minerals within their own narrow channels."

The following table shows that the quantity of Tin Ore smelted in the Colony is increasing rapidly:—

TIN.—The produce of the Colony of New South Wales.

Year.	Tin—Ingots.		Tin Ore.		Total Value.
	Quantity.	Value.	Quantity.	Value.	
	Tons.	£	Tons.	£	£
1872	47	6,482	848	41,221	47,703
1873	904	107,795	3,635	226,641	334,436
1874	4,101	366,189	2,118	118,133	484,322
	5,052	480,466	6,601	385,995	866,461

Mr. Gower, the Mining Registrar at Vegetable Creek, reports:—

It is with considerable regret that I have to announce that nearly all the Tin Fields are suffering from the effects of a most unparallelled drought, which has lasted over four (4) months, thereby causing almost all the mines to suspend operations till sufficient quantities of rain will have fallen, and has also, of course, been productive of a great reduction in the yield of tin ore.

Tin-mining in this district is only in its infancy, as, now that the deposits of tin have been nearly worked out of the bed of Vegetable Creek, the course of which for 3½ miles has proved very rich, the stanniferous wash has been traced in most of the mines into the banks and adjacent flats; which in some instances equals, if not excels, in richness the creek bed, and these discoveries have given increased confidence in the permanency and prosperity of this Division.

The mine of Messrs. Moore & Speare, a property of 160 acres, purchased mineral land, situated on Vegetable Creek, has far surpassed all the others in the yield of tin ore; in fact, it can be reckoned the best paying and richest ground yet worked on the Tin Fields, having declared a dividend of ten thousand pounds (£10,000) for the last twelvemonth's work.

On some parts of the lead, the stripping (a ferruginous sandstone and cement) has been very hard, requiring the use of blasting powder; but on an average, all over the lead, which has proved payable to a width of two hundred (200) feet, the wash is easily got, and the depth of stanniferous dirt is three feet, averaging one bucket (80 lbs.) of tin ore to a load (sixty buckets). The average weekly yield for the last six months, with fifty men and two horses, has been twelve (12) tons. The full complement of men are not at work, on account of the scarcity of water. The amount of tin ore raised since the mine started operations eighteen (18) months ago, is six hundred and twelve (612) tons. There are three large dams constructed, only one of which at present holds sufficient water for use. This valuable mine is ably managed by one of the partners, Mr. Peter Speare.

The next mine of importance, where the richest deposit of stanniferous wash-dirt is now being worked, is the property of the Vegetable Creek Tin Mining Company, situated a quarter of a mile south of the line of main workings along the bed of Vegetable Creek. The lead in this mine has apparently no connection with that worked in the creek, as the wash-dirt is a regular river-bed drift and gravel, while the latter in some places is a coarse creek wash, but is mostly a cement and clay wash, that requires puddling before the tin can be extracted.

A main tunnel has been driven two thousand (2,000) feet along the course of the lead, the width of which is from eighteen (18) feet till it gradually widened out to four hundred feet (where the present workings are), with an average thickness of three feet of excellent paying dirt, and at a depth of sixty (60) feet from the surface. The wash dirt is driven out and raised to surface by means of whips and windlasses, and a tramway is laid to truck the dirt from the shafts to the sluice-box. On one part of the workings, while sinking a shaft to cut the main lead, a very hard layer of cement, fourteen inches thick, was struck, but under it a splendid run of wash was found from two to four feet thick; and sixteen (16) feet below this again the main lead is worked. But the sinking up to the present is entirely through pipe-clay. The country at the 60 feet level is a regular river bed, and in some parts there is fourteen feet of loose drift sand, heavily intermixed with tin ore. In one portion of the lead a tremendous heat or electric current seems to have passed through, as everything is charred, even the tin ore is burnt into clinkers. 600 feet ahead of the Company's present workings, the ground tested by means of boring rods has proved that, after passing through 40 feet of extremely hard rock or cement, and under that a few feet of pipeclay, excellent if not far richer dirt is to be found, proving without a doubt that it is the richest and longest bed of stanniferous dirt ever struck at such a depth in the Colonies.

The managing partner, Mr. J. J. O'Daly, is effecting considerable improvements in his tin-dressing appliances, as he is erecting a counterpart of the machine invented by Mr. Wesley, and working so successfully on the Great Britain Tin Mine. The yield of tin ore from the Vegetable Creek Co's. Mine for the past eighteen months is three hundred and fifty (350) tons. The average weekly yield is seven tons with forty-five men and two horses.

The largest mining property on Vegetable Creek is that of the Great Britain Tin Mining Company, under the management of Mr. W. H. Wesley. It consists of 500 acres, freehold, and situated at the head of the creek. On this mine a tramway was laid by the former management to truck the wash-dirt down to the river Severn, a distance of 6 miles; and a 12-horse power steam-engine erected at the tip heads to pump water for sluicing, but the tramway after being used a few times was found to be an utter failure, thereby incurring a considerable loss to the company, it being found that the dirt could be more economically washed on the mine. The steam-engine is now removed back on the mine, and it is the intention of the present manager to work some new appliances of his own invention for tin-dressing purposes by steam, thereby effecting a considerable saving in horse and manual labour.

This mine, like all the others, is raising but little tin on account of the want of water. There are but half the number of men employed that would otherwise be required if there was a sufficient quantity of water for sluicing. Three large dams have been constructed on this property, one of which, built some time ago, proves of immense value, being the only one full; the two others excavated on another part of the mine, during this long dry weather, when once filled, are capable of holding sufficient water for three months' use. The stanniferous wash-dirt here is not so rich as it is in the lower part of the creek. The alluvial deposit being almost on a level flat, and in consequence of being at the head of the creek, has been less subject to floods; but the tin is more evenly diffused over the width of the lead about two chains (130 feet). The stripping averages about 5 feet deep, and in many places presents an appearance of a ferruginous sandstone. The wash-dirt or stanniferous deposit is about 18 inches deep on a soft claystone reef, and has the appearance of a heap of soft clay with about 50 per cent. of flagstones mixed together and thus allowed to set. The richest deposits occur where the stones are in greatest abundance, and the most cemented. Thus all the alluvial dirt on this mine requires to be well puddled before the tin can be extracted; and for that purpose four large puddling machines have been erected close to the dams and sluicing appliances.

148 tons of tin have been raised from this property during the last fifteen months, and the mine is now in first-rate working order, the only thing retarding the tin-raising is the want of water, and when a sufficient quantity shall have fallen to fill up the large reservoirs, the manager has every hope of doubling that yield during the ensuing twelve months.

The Messrs. Hall, Bros., are also very large holders of mineral land in Vegetable Creek and the vicinity, and have proved themselves most enterprising speculators in prospecting and developing their numerous selections at the Six-mile, the Grampians, Tent Hill, The Springs, and Kangaroo Flat, Strathbogie. The yield of ore from all of them has greatly diminished the last three months, having to suspend almost all operations on account of the prolonged drought. The yield of ore from all the selections of Messrs. Hall, Bros., for the past year is about 400 tons.

At Kangaroo Flat, Strathbogie Run, the tin is raised from a depth of from 20 to 70 feet from surface. The sinking is through a volcanic basalt formation, and the alluvial deposit is a river-bed drift, with occasional patches of cement, but with tin well interspersed.

No portion of the Tin Fields offers stronger inducements than Kangaroo Flat for the investment of capital in developing its resources. In every hill, flat, and gully, for miles around, tin ore can be found in greater or less quantities on the surface, but unfortunately the surface deposits are too poor to pay working men, and capital is required for sinking in search of the lower deposits.

The tin lodes have not yet had much attention directed to or capital invested in them, and even where a fair prospect of testing the lodes with advantage presents itself, the owners of the land ask such exorbitant prices as to deter capitalists from investing; thus this branch of tin mining, upon which the future and permanent prosperity of this district must mainly depend, is at present lying dormant.

The only property taken up for the purpose of working lode tin and on which a quantity of work has been done by an English company, under the management of Captain Stevens, is situated at the Nine-mile Mole Tableland.

A battery of eight stampers is being erected on this mine, there are a thousand tons of quartz at grass ready to be operated on, and in respect of which many persons well acquainted with the locality speak in the most sanguine terms.

The great disadvantage under which all the tin districts suffer is the absence of a permanent supply of water, preventing sluicing operations being carried on with regularity on an extensive scale.

Most of the principal operations in tin-mining are carried on in wide open creeks or gullies and flats; hence every mineral leaseholder has to incur a rather serious expense in either erecting a strong dam or excavating a sufficiently large space for water storage.

The extent of tin-bearing country is something enormous, and in many places the percentage of tin is so small that a considerable area will remain unwrought, unless a far more economical and quicker system of extracting the tin from the sand and gravel can be invented.

The necessity for economy in the use of water, and in the employment of manual labor has stimulated some of the more enterprising and skilful miners in this district to introduce appliances which will reduce the consumption of water and the number of men employed, and at the same time increase the quantity of such dirt treated in a given time.

The first system of washing stanniferous dirt was in the usual style of sluicing for gold, using the narrow box and sluice fork; but after the rich deposits of tin were washed in this manner, the miners found that a good deal of their mineral ground would not pay wages unless some other means were found to wash the dirt quicker and cheaper.

The hopper plate and large sluice box were then introduced. These hopper plates are now in general use, and are 4 feet long by 2 feet wide, perforated with holes from $\frac{3}{8}$ to 1 inch in diameter, and fitted at the head of the sluice box. This has proved to be a considerable improvement on the original manner of sluicing, for a larger quantity of puddled dirt can be washed per day, the principal reason being that the stones or pebbles are not allowed to drop into the sluice box; the fine stanniferous dirt is in consequence far more easily washed, a square mouthed shovel being used in place of the fork to keep the tin up to the head of the box.

Another great saving in the cost of tin sluicing was effected by substituting horse power for that of manual labour in pumping water; for where it used to cost 12s. to 14s. per day for two youths' labour at the Californian pump, the only expense to the miner by the present system is the forage for one horse, reckoned at 4s. per day. This horse pumping gear is now in general use in most of the mines of any note, and is found to work capitally, not only on account of economy, but because of the steady stream of water thrown, which is most essential for sluicing purposes. By this present system, the cost of washing does not exceed 2s. for every load or ton of puddled dirt; whereas by the former process the expense per load would be from 2s. 6d. to 3s.

By the energy and perseverance of Mr. W. H. Wesley, Manager of the Great Britain Tin Mining Company, the horse pumping gear and hopper plate have been vastly improved upon.

Mr. Wesley's apparatus, a drawing of which is annexed, consists of a horse engine and rope belting, which works a Californian pump, 18 feet long and 9 inches by 3 inches; the vertical height of water raised being 11 feet. On the same shaft which works the pump is a wheel and belt, which works a cylindrical sieve 1 foot 8 inches in diameter by 3 feet long, made with bars of iron ½ inch square placed longitudinally, and a space between each bar of ¼ inch, so that any stones or lumps over ¼ inch diameter cannot go between the bars, and consequently do not pass into the sluice box.

On each end of this sieve flanges are riveted at the receiving end of the sieve; this flange is to prevent the dirt from dropping out underneath the hopper into which the wash dirt is shovelled and prevent its falling into the sluice box. The flange at the discharging end is to prevent the stones from escaping too quickly, so that they have a better chance of being washed clean, and also tends to stop all the water that might otherwise run longitudinally over the sieve and go away with the stones or pebbles, carrying some loose tin with it.

This cylindrical sieve is fixed over the head of the sluice-box, and set at an angle of 3 inches in its whole length, to facilitate its discharging the stones or pebbles, which are washed clean in a few revolutions. At the back of this sieve the water is conveyed by launders from the pump to the hopper or receptacle for feeding. The stanniferous dirt is thrown into this receptacle, through which three-fourths of the water used for sluicing passes, and carries the wash-dirt into the sieve. One-fourth of the water is conveyed through pipes or tubes from the launder to bear direct against the pebbles revolving against the streams of water with the revolution of the sieve, consequently these streams of water falling on these pebbles whilst in motion allow every chance to wash off all the loose tin. The pebbles pass through the sieve and fall into a bin with a sloping bottom, to which a trap-door is fixed worked with a lever, under which a truck (3 feet 6 inches x 3 feet 6 inches x 1 foot 9 inches) is placed and a tramway is laid. The trap-door being raised the pebbles fall into this truck, which is removed by a horse a distance of 100 yards, and there discharged to fill up old worked ground. The fine stanniferous sand and gravel fall through the sieve into the sluice box (18 feet long, 3 feet wide at head, by 1 foot 6 inches at tail) on which two men using square-mouthed shovels wash the dirt, in a similar manner to the old way of sluicing with forks. The refuse or tailings from the sluice-box fall into a hutch (see drawing), under which a truck road is laid, and the truck (of the same dimensions as the one given above for pebbles) goes underneath this hutch, in the bottom of which are three valves or slides 4 inches x 4 inches, worked by a lever, which when opened, the pressure of water forces the refuse and tailings into the truck, and fills it in a very short time; the truck is then drawn away by a horse to a suitable distance out of the way of any of the workings.

Two puddling machines are supplied with water from the hutch at the tail of the sluice-box, and all the surplus water is conveyed back into the dam by a launder. The quantity of puddled dirt washed per day by the above machinery is not less than sixty-five tons by the following number of men:—

	s.	d.
Two men feeding (shovelling wash-dirt into sieve) at 8s. each	16	0
One man at head of sluice-box	8	0
One youth behind do.	5	0
One horse removing tailings by truck	3	4
One youth driving do.	5	6
One horse removing hopperings, pebbles, by truck	3	4
One youth driving do.	5	6
One horse working engine, &c.	3	4
	50	0

One man's labour must now be deducted from the above calculation, for pumping water for the two puddling machines = 7s. 6d., which leaves the cost of washing sixty-five tons of puddled dirt per day at 42s. 6d., or less than eight-pence per ton.

As Mr. Wesley has most kindly furnished me with a drawing of his tin-dressing machinery, I hereby take the opportunity of forwarding you the same, which will enable you to understand more fully the worth of his invention, for it has proved to be a most unqualified success. By it, three times the quantity of puddled dirt can be washed per day, at a more reasonable rate than by any system of sluicing or tin-extracting now being used in the tin mines.

This machinery is within the reach of any party of working miners, as it can be erected at a cost of £60 to £70, and is so extremely simple that it would be very soon understood; and will prove of immense benefit not only towards developing the district, but in enabling the tin miners to make poor ground pay very remunerative wages. The excellence of this invention is being recognized on this creek, as the Vegetable Creek Tin Mining Co. are erecting one on the same principle as that now being used so successfully at the Great Britain Tin Mining Company's mine.

Messrs. Moffatt and Harridge, of Stanthorpe, are erecting a Tin-smelting Works at Tent Hill, four miles from Vegetable Creek, and they expect to be in full working order in March next, which no doubt will prove of great benefit to the miners, as at present all the ore raised in this District is sent to Newcastle or Sydney Smelting Works, thereby incurring a serious expense in carting. The number of mines of any note, raising tin ore in this Division are twenty-five (25), the total yield of which for the past year was 2,080 tons of ore. Of course there are a great number of small tribute parties working mineral ground, of which I could not get any reliable information, but I believe all who are so engaged are making a very comfortable living. There are 500 miners employed in this Division.

The discovery of new rich deep leads, and the opening up of fresh mineral ground in this district, will give fresh impetus to mining enterprise, and altogether I look forward with confidence to the year 1875 proving a prosperous one for the Tin Mines of New England.

RETURN of the Quantity and Value of Tin Ore raised from the undermentioned Tin Mines at Vegetable Creek.

Name of Mine.	Where situated.	Quantity produced and Value.	
		Quantity.	Value.
		Tons.	£
Moore & Speare's	Vegetable Creek.	410	39,360
Baalgammon Claim		120	11,520
Little Wonder		65	6,240
Cubiss (Tribute)		100	9,600
Little Britain		120	11,520
Nonpareil (Marshall)		50	4,800
Tornisson's		50	4,800
Vegetable Creek Mine (Hall Bros.)		120	11,529
Grampians		100	9,600
Six-mile		80	7,680
Tent Hill		60	5,760
Kangaroo Flat		30	2,880
Springs		10	960
Rose Valley		100	9,600
Great Britain		120	11,520
Vegetable Creek Company		234	22,464
Water-holes		50	4,800
Glen Creek		40	3,840
Banca		30	2,880
Tent Hill (Arnott's)		40	3,840
Six-mile Company		80	7,680
Gulf Creek		50	4,800
		2,059	£197,304

The following is a report furnished by Mr. Wilkinson (the Geological Surveyor of this Department), to the Surveyor General, in July, 1873, upon the Tin-bearing country of New England:—

I HAVE the honor to inform you, for the information of the Honorable the Minister for Lands, that I have further examined the Cope's Creek and other tin-bearing localities in the district of Inverell; and, in accordance with your instructions, I now forward you the following additional observations on the geology of this interesting and important district.

Dr. Ludwig Leichhardt, as you are aware, in 1842–43, made a cursory exploration of the western part of New England, and gave a short description of its geological features; but in 1853, the Rev. W. B. Clarke made a more extensive examination of the whole district, and, in his report (dated 7 May, 1853) to the Honorable the Colonial Secretary, drew attention for the first time to the probable occurrence of extensive deposits of tin ore. He then stated that "wolfram and oxide of tin, with tourmaline, occur near Dundee and in Paradise Creek, and it is probable that this ore of tin is plentifully distributed in the alluvia of other tracts, as I have found it amidst the spinelle rubies, oriental emeralds, sapphires, and other gems of the detritus from granite." That these anticipations have been realized is attested by those valuable deposits of tin ore, which have now been proved such an important addition to the vast mineral resources of New South Wales.

As, however, the Rev. Mr. Clarke, in his report to which I shall hereafter make frequent reference, dwells more particularly on the leading geological features of the Northern Districts, I will do myself the honor, having the benefit of his previous researches, to give you a more detailed description of that part of the tin-bearing country which lies chiefly to the south and east, and within a radius of about 25 miles from Inverell.

The principal tin mines within this area are those on Cope's Creek, Middle Creek, and on the Macintyre River at Elsmore and Newstead.

From Newstead the Macintyre River has a westerly course to Inverell,—distance about 12 miles; it is then diverted in a northerly direction by the high basaltic range of Table Top. Middle Creek, coming from the S.E., flows into the Macintyre, a short distance above Inverell; whilst Cope's Creek, lying about 10 miles further south, takes a westerly course and joins the Gwydir or Big River.

The general aspect of the intervening country is very uneven and rough, consisting of rugged hills more or less thickly timbered, and rocky gullies and creeks, which in places have their channels cut into wild precipitous ravines, as on Lower Cope's Creek, where also several fine waterfalls may be seen.

By observations with an aneroid barometer, the township of Inverell is about 2,010 feet above the level of the sea; and the vicinity of Cope's Creek is from 300 to 800 feet higher.

For clearness of description, it may be well to arrange the formations separately in the following order:—

Recent	Quaternary.
Pleistocene	Tertiary.
Pliocene...	
Miocene...	
Carboniferous	Primary.
Granites...	
Greenstone	

I would here premise that, in using the above terms pleistocene, pliocene, &c., I do so in a measure provisionally, as expressing the *relative* ages of the formations to which they are applied; for, in the absence of fossil evidence, as the Rev. Mr. Clarke has pointed out (in his *Remarks on the Sedimentary Formations of N. S. Wales*), the exact age of some of the N. S. Wales tertiary deposits cannot be definitely ascertained. Nevertheless, as those I have examined in the Inverell District correlate, both in lithological character and in their relation to the physical geology of the country, with formations of the above-mentioned ages in Victoria, where the tertiary divisions are more clearly defined, and with which I am personally familiar, the use of the above terms may be justified here.

Sketch section showing relative positions of the formations:—*a*, recent river deposits. *b*, pleistocene terrace drifts along river valley. *c*, newer pliocene "lead," covered by basaltic trap. *d e*, older tertiary tin-bearing gravels, clays, and ironstone, containing lower miocene fossils, leaves, and plant-stems. *f*, carboniferous strata, upheaved by granite. *g h*, greenstone trap, penetrated by granite dykes.

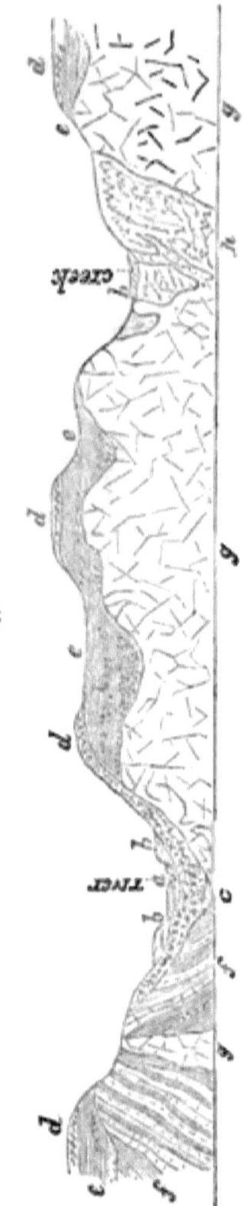

Fig. 2.

RECENT.

The deposits under this head are too well known to require much comment. They embrace all those river drifts, alluvial and other surface accumulations, which are in course of formation at the present time. They are of great economic importance both to the agriculturist and to the miner. In the Coinet Tin Mine, Cope's Creek, eight men have washed out 6 to 8 cwt. of stream tin per day.

Further up Cope's Creek, in the Lyngarra Mine at Captain Swinton's Station, and in the adjoining Victoria Tin Mine, even richer yields have been obtained.

This locality, I believe, contains the richest and most permanent of the *recent* tin deposits in the district. On Middle Creek, about 30 tons of ore were obtained by fifteen men in three months; well cleaned ore is worth at the mine from £50 to £60 per ton. Other instances might also be given to show the value of the *recent* alluvia. They are

Sluicing for Tin-ore, Britannia Tin Mine.
COPE'S CREEK, NEW ENGLAND

perhaps the most easily worked of all the tin deposits; but occurring as they do, along the creek-beds, the mining operations are very likely to be impeded by floods, of which miners have had discouraging experience during the past summer months.

Gold has been found, though not in sufficient quantity to pay, in all the tin-bearing deposits. Sapphires are of common occurrence; some of them are of large size and good colour, and worth up to £15 or £20; the miners, however, pay but little attention to the saving of them.

As included with the *recent* accumulations, may also be mentioned the frequent additions of muddy sediment deposited over the river flats and other low-lying lands by floods. This may be realized from the effects of the late heavy flood at Inverell, which bear testimony as to the thickness of the sediment left on the floors of the houses and on the river flats after but one inundation. The enormous amount of earthy matter thus annually brought down by streams and redeposited is very apparent.

Whilst these accumulations are taking place, other effects of denudation may be noticed. I refer to those deep dykes or gullies now furrowing the sides of hills and cutting through the alluvial flats. They may be well seen on the river flats near Inverell: where, twenty years ago, the rain-water would spread out and flow away over the unbroken surface of the ground, it has since eroded channels, 10 or 15 feet deep, which had their origin in the narrow gullies formed by dray-tracks and cattle-pads. To what extent these newly-formed drainage channels, by the greater facilities they afford for the rain-water to run off, may increase the liability of the rivers and creeks to be flooded in the future, is a subject not unworthy of some consideration.

Pleistocene.

The pleistocene formation includes those drift deposits forming alluvial flats which are found more or less in all the valleys, and through which most of the present streams have worn their channels.

They consist of gravel, sand, clays, and loam, varying in arrangement, and their composition depending very much on the nature of the rocks from which they have been derived. Thus, in granite country, the detritus is of a coarse sandy character, with a little quartz drift; that from the older tertiary formation consists chiefly of water-worn gravel and sandy ferruginous clays; from the basaltic trap have resulted thick deposits of black and red loamy clay, affording a very fertile soil; and those extensive alluvial flats along the Macintyre River are formed of the detritus from all these rocks, together with that brought down by the river from other formations in distant localities. In the valley of the Macintyre, as in the river valleys in other parts of the Colony, several of these alluvial deposits occur at different heights, forming terraces on the sides of the valleys. (See fig. 2. *b*.) One patch of this drift, consisting of large water-worn boulders and pebbles, may be seen on the south bank of the river near Inverell; it rests on basalt, at about 40 feet above the bed of the river, and is now out of reach of floods. These terrace drifts, therefore, mark the succesive levels of the valley as it became gradually scooped out and deepened by the action of the drainage water flowing down from the high ranges of the Cordillera.

The Rev. Mr. Clarke, referring to these geological features, remarks :—Looking to the Colony of New South Wales, we find that in more than one instance the present river channels have deepened since the drift first began to crowd their banks. I have traced one of those drifts streams, sometimes at great heights above the valleys, for more than 80 miles."

The river flats form rich agricultural land, and have been taken advantage of for that purpose.

In the creeks and gullies traversing the granite country, the alluvial deposits are all tin-bearing ; but, being often of considerable thickness—sometimes 20 feet—they are not so easily worked, on account of the great amount of stripping required, as the shallower and more recent drift along the beds of the creeks.

At the Lady Emily Tin Mine, Cope's Creek, the pleistocene drift, consisting of coarse rounded drift and yellow sandy clay, is about 16 feet thick ; the wash-dirt varies up to 5 feet thick, and from it as much as 4 lbs of stream tin to the dish have been obtained.

The following is a sketch section across Cope's Creek, near this mine :—

Fig. 4.

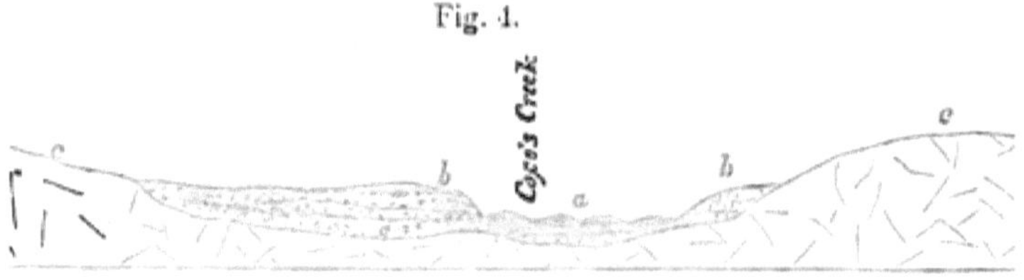

a, recent alluvial, 8 feet thick, with 2 feet of wash-dirt. *b*, pleistocene drift, 16 feet thick—from 1 to 5 feet wash-dirt. *c*, granite.

As to the permanency of the yield of tin ore from these alluvial deposits it is impossible to speak with certainty ; but even a cursory examination of Cope's Creek and its vicinity cannot but convince one that the tin-bearing ground as yet unworked will afford continuous and profitable employment to miners for some years.

Besides the alluvia along the creeks, several rich patches of surfacing from 1 to 4 feet thick have in various places been opened, the wash-dirt requiring but little "stripping," and yielding from 15 to 30 lbs. of tin ore to the cart-load of dirt. Messrs. Ross, Martin, and Irwin's mine, near Captain Swinton's station, Cope's Creek, and Messrs. Reeves & Co.'s mine, Long Gully, are instances. The richness of these surface deposits, together with the angular form of the tin ore, often indicates the proximity of lodes in the underlying granite, to the breaking up of which the supply of tin ore is due. Owing to these numerous tin veins or lodes, the surface soil, where derived from the granite, has been found almost everywhere to contain tin ore. We may reasonably believe, therefore, that these rich patches of surfacing already opened are but a few amongst the many that will yet be discovered and profitably worked.

About 2 miles above Mrs. Anderson's station, on Newstead Creek, a deposit of tufaceous limestone occurs: it is 4 feet thick, and rests on calcareous clay, and lies at the junction of trap with carboniferous rocks. Springs issuing from those rocks are evidently the origin of this limestone. It is now being quarried and burned, and yields lime of fair quality.

PLIOCENE.

Next in order of sequence is the basaltic trap. For the miner this volcanic rock has but little interest, but to its influence the best pastoral and agricultural land in the district chiefly owes its fertility; a reflection that may afford some consolation to those who in wet weather deplore the existence of the notorious "black soil" of Inverell, for, as before remarked, it is the disintegration of the trap that has produced this fertile soil.

Around the township of Inverell, and in places throughout the whole district, the basalt formation occurs. It extends for some distance to the westward; and to the south-west it forms the watershed between the Gwydir and Macintyre Rivers

The basalt varies greatly in thickness. At Inverell it forms the bed of the Macintyre, and on the west bank attains a thickness of several hundred feet; while a short distance to the eastward, on the Newstead Road, it occurs only a few feet thick, capping a hill about 200 feet above the level of the river. Many other similar instances might be mentioned which mark the uneven surface of the ground at the period of the volcanic eruption, when a flood as it were of molten basalt overflowed the country. Within the district I have examined I have not been able to determine any of the points of eruption whence this lava issued, unless the tufaceous trap of Table Top may perhaps indicate that hill as one, but this is doubtful.

From what the Rev. Mr. Clarke's report states, it appears that the volcanic vents lie to the eastward. The trappean rocks are described as "bursting through both granite and porphyry and overflowing them; they form the culminating points of the Cordillera on the Ben Lomond Range, and break out along the spurs from that range in various places on the western falls. That they have issued from the granite is shown very remarkably by several examples along the banks of the Macintyre, a little below the junction of Ouerra Creek, and upon the broken ranges between the head of Paradise Creek and the junction with the river."

Near the junction of Newstead Creek with the Macintyre the basalt appears to have filled up an old valley, and the same features are noticeable further down the river, near Brodie's Plains, and again at Inverell. An examination of the country to the eastward may possibly show this old valley to have been in pliocene times that of the Macintyre, down which the basaltic lava stream poured, and damming it up, caused the drainage water to erode a fresh channel, forming the existing valley, which during the succeeding pleistocene and recent times has not been deepened to the level of the old valley. (See fig. 2, *a c*.)

These geological features are met with in several of the newer pliocene leads in Victoria, especially in the Durham lead, near Ballarat, where the lava has flowed down a valley and covered up the bed of the old stream or lead with a thickness of nearly 300 feet of basalt. In New South Wales similar auriferous leads occur at Lucknow, Gulgong, &c. As, therefore, these leads traversing gold-bearing formations have been found highly auriferous, it is not improbable that the Newstead one may contain payable stream tin, provided, of course, that it has passed through stanniferous granite country; however, I cannot speak, my examination not having extended further to the eastward.

Another small lead, covered with basalt, occurs between Middle Creek and the Macintyre; it runs in a north-westerly direction, between two granite ranges, until it enters the Macintyre valley. This small lead has not yet been prospected, and I believe it to be well worth the attention of the miner.

An interesting cliff section of basalt may be seen on Mr. Colin Ross's property on the bank of the river at Inverell. The following is a sketch of it:—

Fig. 5.

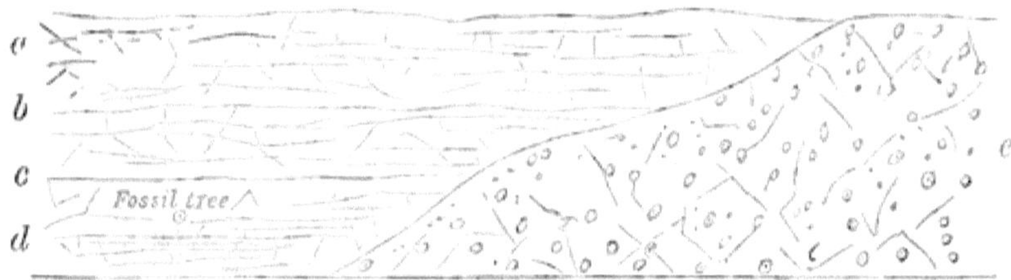

a b, amygdaloidal basalt, much decomposed. *c*, friable cellular basalt, enclosing fragments of wood and pieces of earth. *d*, dense columnar basalt. *e*, volcanic breccia, composed of fragments of basalt of various sizes, embedded in an indurated volcanic mud, much stained with peroxide of iron, which imparts to the rock varying shades of deep red and yellow. This breccia is older than the *a b c d*, and evidently formed the side of a hill on which plants were growing at the time of the basalt eruption; for at the junction of the basalt and breccia lies a thin bed of red clay, the former surface soil, in which I discovered numerous stems of plants. Some of these stems are in an upright position, and even penetrate a few inches into the basalt rock above, and several I found with the woody matter but little altered. These facts are very singular, as proving the viscid state of the overflowing basaltic lava, to have thus surrounded the small plants without destroying them, and how rapidly it must have cooled. Another interesting relic of the newer pliocene period that this section reveals is the trunk of a tree, about 2 feet in diameter imbedded in the layer of basalt marked *c* in the above sketch.

The wood, though much changed, yet retains its fibrous structure most completely. It somewhat resembles the stringy-bark, and may possibly be a species of *Eucalyptus*; but this is difficult to decide without the aid of the microscope.

Surrounding the tree is a soft substance, 2 inches thick, which was probably the bark. Small masses of yellowish earth are also scattered through the same layer of basalt. The rock above this is full of amygdaloidal cavities, containing large double hexagonal pyramids of that rare zeolite, *Herschelite*, together with calcite and minute crystals of analcime. I have also seen large radiating crystals, several inches in length, of arragonite from the trap of this locality.

The columnar structure which basalt frequently assumes may be observed in the rock marked *a* in the above sketch. The basalt has here rudely crystallized in vertical hexagonal and pentagonal columns; indeed the flat surface of the rock across the river bed has the appearance of a roughly formed pavement of five and six-sided blocks of stone.

In another cliff section, about 2 miles farther up the river, the basalt has a radiated columnar structure, the narrow columns radiating like the ribs of an open fan, and giving to the cliff a very picturesque appearance.

At Newstead Station the basalt contains abundance of olivine, in small roundish masses, composed of an aggregate of angular grains of red, yellow, and green colours. Veins of white jasper, several inches thick, are also not uncommon in this rock.

Near Auburn Vale Station the basalt is of a trachytic character.

It has been before remarked that outliers of the volcanic trap occur in places throughout the district. These outlying vestiges of a once overspreading formation now testify of the vast denudation the land has undergone since the pliocene period. (See fig. 2, *d d*.)

Miocene.

The rocks of this period are of much economic importance, on account of the valuable deposits of stream tin which some of them contain.

Certain of these rocks are full of impressions of leaves and plant stems, which are believed to be of lower miocene age; but I shall refer to them hereafter.

The formation must have been of considerable thickness, and seems to have once covered nearly the whole district, for it is found on the summits of hills, and again filling some of the intervening depressions. The upper part consists of sandy concretionary ironstone, which sometimes assumes a pisolitic structure, from the small ferruginous concretions composing the mass. Where this is the case, the surface of the ground is often strewn with small round ironstone pebbles of the size of a pea. The ironstone is now chiefly found in outliers forming those "red hills," as they are locally called, which constitute such marked features in the country between the Macintyre and Cope's Creek.

Underlying the ironstone are red and white sandy clays, and beneath these waterworn drift and conglomerates, the latter usually occupying the sides and bottoms of the ancient valleys, and containing the tin-bearing deposits.

Owing to the enormous denudation this formation has suffered, the overlying clays and ironstones have in places been entirely removed, leaving the stanniferous gravels exposed near the present surface, and therefore easily accessible to the miner. Such is the case at the Elsmore, Stannifer, and other important tin mines.

At the Elsmore Mine this old alluvial drift has been broken up and redeposited, forming shallow surfacing near the top of a hill, where it is now being worked; but in another part it lies undisturbed, and consists of a hard conglomerate, the water-worn boulders being cemented together with a siliceous cement.

The hill on which the drift occurs formed the south side of one of the above-mentioned old valleys; the drift therefore deepens as it is followed to a lower level, and there is every probability that the wash-dirt will not decrease in richness, though the amount of stripping will be greater. The same description of conglomerate as that at Elsmore is to be met with again about 3 miles to the eastward, at the Karaula Tin Mine; it is rich in very waterworn stream tin, and I have seen specimens of it also enclosing coarse specks of gold with the tin ore.

The surfacing at the Stannifer Tin Mine, Middle Creek, is the remains of an old lead, through which the present valley has been eroded, thus leaving the gravelly bed of the old stream now in disconnected patches amidst the hills. Water-worn pebbles and boulders, up to 1 foot or more in diameter, of quartz and quartzite, compose the drift, which in places is cemented into a hard ironstone grit and conglomerate. But little striping is required, and the wash-stuff is carted for about ¾ a mile down to the creek and then sluiced. About 4,000 cubic yards yielded 57 tons of stream tin. Small crystals of chromic iron and one small diamond were obtained with the tin ore. I have traced this lead for about 2 miles to the westward, where it has been opened by Messrs. Burlington & Low. It has here yielded from 10 to 50 lbs. of stream tin to the load of wash-dirt. Waterworn pebbles of clear transparent and dark smoky-coloured crystal quartz are very abundant. The drift is sometimes changed into a hard silicious and ferruginous conglomerate.

For some distance further to the westward this lead has been removed by denudation; but the red hills on the S.W. side of Middle Creek suggested its continuance in that direction and towards the Ponds Creek.

I believe that, to the fact of the present Middle Creek having cut through the old lead and redistributed its rich contents, is to be attributed the richness of the more recent alluvia in the Sydney and Ancient Briton tin mines, about 2 miles lower down the creek.

Near the head of the Ponds Creek is another similar old lead, and which I believe will become an important one; it has been partially prospected, and with good results. Several topazes of bluish-white colour were also obtained from it.

On the Boundary Tin Min, a few miles further to the eastward, lying between two granite ranges, occurs an older basalt, of a brown and purplish colour, and much decomposed. It may possibly have some relation to the volcanic breccia underlying the pliocene basalt in the section at Inverell, and to another similar breccia exposed in a cliff section on Newstead Creek. (See fig. 6, *e*.) In appearance this decomposed basalt resembles some of the older basalts which are interpolated with miocene strata near Geelong, in Victoria; it may, therefore, be of the same age. It seems to lie in an old valley, and has probably covered up a lead which may be the continuation of the one just described near the head of Ponds Creek. Mr. David Wilson, Manager of the Boundary Mine, had a prospecting shaft put down about 100 feet through the basalt, bottoming on granite, without any intervening drift; but as the bed rock was dipping fast to the north the deeper ground will be found in that direction, and which the conformation of the contiguous granite ranges would also suggest.

In my former report I mentioned an old lead which crosses the New Banca Tin-mining Company's ground. I have since traced this lead for more than a mile to the south-west; its course being marked in places by a hard, white, silicious conglomerate. I believe that it crosses Cope's Creek, between Captain Swinton's station and Tiengha, and that the tract of country occupied by tertiary red sandy clay, south of Cope's Creek and passing near Stanborough, indicates the course it has taken. I cannot, however, speak with certainty on this point, not having had sufficient time to examine the ground minutely.

Outlying patches of drift, partly covered by basaltic trap, occur on the hills south of Cope's Creek, near Sutherland's Water, and also on the Bismarck Tin Mine.

Between Inverell and Middleton the tertiary rocks forming the red hills cover up the granite formation to a considerable extent. I am not aware of its having been prospected. There may, however, exist as payable drift in this as in those localities above described.

On the Borah Creek, which flows into the Gwydir, about 2 miles above the junction of the river with Cope's Creek, several patches of pebble-drift, iron-stone, and clays, capped with basalt, mark the course of a small lead. Not far from its source the Borah Creek crosses the lead, and for about 2½ miles further down the valley, which lies between abrupt granite ranges, it has been entirely denuded; but below this point it may be traced now and again, in a N.E. direction, by patches of drift covered with basalt. It is immediately below where the creek has first cut through the lead that the operations of the Borah Tin and Diamond Mining Company have been carried on in sluicing the more recent alluvial drift. Beside several tons of stream tin, upwards of 200 diamonds were obtained in a few months. Mr. Thos. Adams, one of the proprietors, showed eighty-six of the diamonds, which I weighed and found to average one carat grain each —the largest of them weighing 5·5 carat grains. They were mostly of a light-straw and pale-greenish colour. Several were nearly octahedral crystals, but the rest were modifications, with curved facets and edges, some appearing almost spherical in shape. Sapphires and garnets occur in the diamond-bearing drift, together with small polished black pebbles. If the diamonds have been derived from the old lead, no doubt many more will be found where it has been entirely denuded, and the tin ore and diamonds it contained have been redistributed in the alluvial deposits lower down the creek. From the Bengonover Mine (Messrs. Butler, Swansons, & Co.), about 2 miles below the Borah Mine, I examined several diamonds; the largest, not of good form, weighed 7·5 carat grains, and gave specific gravity 3·4.

At the time of my visit the diamonds were reported from the Ruby Tin Mine on the Borah Creek, about a mile above its junction with the Gwydir River.

The fact that the diamonds from the Borah Mine are found in the creek, immediately below where it has cut through the lead, suggests the impression that they have been derived from the lead. The small black pebbles associated with the diamonds may point to the former existence of another formation, but no vestige of it *in situ*, that I am aware of, now remains to prove it. The surrounding country appears to be entirely of granite. Whether the lead may be the original matrix of the diamonds is a question difficult to determine; it may, however, be mentioned that the facets and edges of the diamond crystals do not appear to be in the least degree waterworn or abraded. I have been shown two diamonds said to have been found near Newstead. Another one, as I before stated, was obtained with the tin ore from the old tertiary drift at the Stannifer Tin Mine, Middle Creek, and three others have been discovered in Darby's branch creek, at the Britannia Tin Mine.

Carboniferous.

On Newstead Station, thin bedded shales of bluish-grey and yellow colour crop out, dipping at an angle of 15° in one place and almost vertical in others, with a general northerly strike. I could not detect any fossils in them, but from their lithological character there is little doubt but that they form part of the carboniferous formation of which the Rev. Mr. Clarke's report states "that the middle beds of this formation, those of the Hunter and Hawkesbury, are widely distributed on the western border of the country between New England and the interior."

A good section, most interesting as throwing some additional light on the probable age of the tin-bearing granites to which I shall hereafter refer, may be seen on Newstead Creek, about 1 mile above Mrs. Anderson's residence. (See fig. 6.)

Fig. 6.—Sketch Section on Newstead Creek.

a, thin bedded grey and yellow shales. *b*, coarse-grained porphyritic granite. *c*, red ferruginous volcanic breccia, similar to that underlying the basalt at Inverell. (See fig. 5, *c*.) This section shows the extension of the granite to have highly inclined the carboniferous strata. For some distance on either side of the granite the shales are much indurated. Carbonate of iron (*sphærosiderite*) occurs, lining the crevices in exposed shales.

Some hard sandstones, probably carboniferous, crop out in the banks of the Macintyre, about a mile below Inverell.

To the north-west the carboniferous formation appears to be extensively developed. The Rev. Mr. Clarke thus describes it:—"The neighbourhood of Warialda furnishes the best position for examining the geological phenomenon connected with the succession of the carboniferous and the underlying formations. From the head of Reedy Creek, which rises near Coragin, to its junction with the Gwydir, there are many instructive superpositions of strata.

"At that place the porphyry is covered by beds of conglomerate and sandstone, which contain seams of cannel coal that have occasionally been used in furnace; the conglomerates are coarse, and pass into sandstone as on the Hunter.

"These are surmounted by beds of grit and sandstone and ferruginous conglomerate, which alternate together, the whole presenting a series of beds which, in colour consistency, and all physical conditions of structure, are in no degree different from beds of the same formation which I have explored in various parts of the sea-board. Over the conglomerates of Reedy Creek, sandstones quite undistinguishable from those of the great sandstone territory of New South Wales rise in succession, till the formation attains a height of from 1,300 to 1,800 feet above the sea, except where it caps the range opposite Bingera, that point being about 400 feet higher.

"I do not doubt that coal in some abundance will be found in the range of those beds, the strike of which seems to be north-east, as on the coast."

Granite.

The granite formation is of special interest, chiefly on account of its tin lodes, and from its having been the original source of the stream tin.

There are granites of at least two eruptions; these may be well distinguished at Captain Swinton's station, where the following sketch plan is taken:—

Fig. 7.

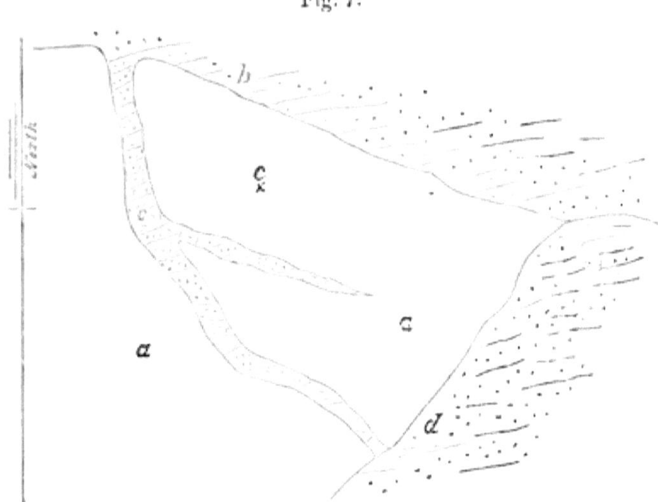

a, hard, dense greenstone trap. *b*, granite, chiefly of euritic character. *c*, eurite dyke traversing the greenstone. This dyke is about 10 yards wide, and runs in a curving manner in a S.E. direction, until it is abruptly cut off by a white porphyritic granite, *d*, full of rectangular crystals of white orthoclase felspar, about an inch in length, which stand prominently out on the exposed surface of the rock. This white porphyritic granite weathers into round-shaped bosses, whereas the finer-grained eurite granite, *b*, is generally broken up into angular masses.

The dyke *c* is crossed by joints bearing E. 5° N.; it is composed of fine-grained eurite, containing a little mica, and irregular veins of quartz and patches of quartz and felspar; it is of much the same composition and structure as the granite *b*, and seems to branch from it.

At the point marked *e* in the sketch the greenstone is traversed in all directions by thin veins of eurite, from a mere thread up to 4 inches in thickness. It is evident, therefore, that the granite *b*, with its offshoot *c*, is younger than the greenstone, *a*; and also, that the porphyritic granite, *d*, is of still later formation.

Both these granites are affected by the same systems of joints which I have already described in my first report.

Throughout the district the granites greatly change in structure and composition; the constituent minerals—quartz, felspar, and mica—presenting an almost infinite variety of admixture and colour.

Quartz veins, occasionally several feet thick, as at the Rex Tin Mine, Middle Creek, are frequent in the granite. The general strike of these is between E. 5° N. and N.E., while others have a meridional strike. The latter are often much broken by "faults." I give sketches of two instances. Fig. 8 is of one near the Cope Hardinge Mine, and fig. 9 near the Rex Tin Mine.

Fig. 8. Fig. 9.

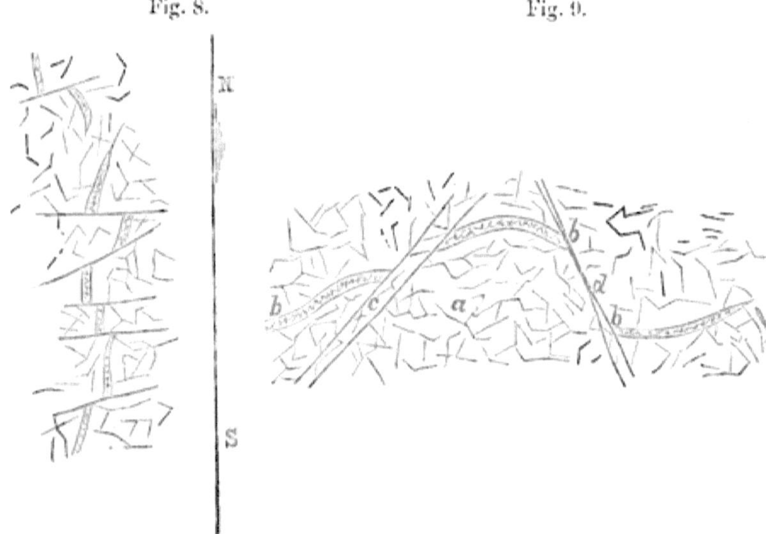

Quartz veins ½ inch thick, faulted by joints traversing ordinary granite with black mica.

a, ordinary granite with black mica. *b b*, vein of quartzite, 4 inches thick, faulted by joints, *c d*, bearing N.E. and N. 25° W.

At the Butchart Tin Mine, next to the Inverell Tin Mine, near Cope's Creek, a quantity of fine lumps of solid tin ore were obtained from the cap of a load. One of these pieces weighed 57 lbs. The ore is of the black variety, and, excepting a little quartz, is very free from foreign matter. An assay of it by C. Watt, of Sydney, gave 76 per cent. tin.

The tin ore at the Bolitho Mine (generally called "Simoni's") runs in irregular veins through a felspathic dyke about 18 inches wide, and dipping nearly vertical. Occasionally these veins of ore unite, forming an almost solid mass of ore, and again thin out and are lost for a time. Several large blocks of the veinstone—one weighing nearly half a ton—were raised, the greater part of which consisted of tin ore. These large specimens may now be seen, I believe, at the offices of Messrs. Beilby & Scott, Sydney.

A vein of solid ore, 4 inches thick, has been opened in the Boundary Tin Mine. Fluor spar occurs in this lode, and also a greenish-yellow steatitic clay.

A dyke of euritic granite, bearing E. 15° N., containing tin veins, has been discovered at the Bismarck Tin Mine, south of Cope's Creek. The tin ore is associated with quartz veins, from a mere string to 3 inches thick, and traversing the dyke in various directions, forming a sort of network of veins.

The ore is usually crystallized in square prisms, lining the sides of the fissure, with the quartz filling the centre; sometimes, however, the whole vein makes into quartz, with separate tin crystals scattered through it.

Mr. Bush, the assayer at Tiengha, kindly gave me some fine crystals of quartz which he obtained from the Albion Tin Mine, Cope's Creek. These crystals are studded on the outside as well as within with black crystals of *cassiterite*, suggesting that the tin and silica were held in solution, and that they both crystallized therefrom at the same time. Specimens of these I have placed in your Geological Museum.

About 2 miles N.E. from Captain Swinton's station are several small tin lodes, associated with veins of quartz and eurite, and traversing in an E.N.E. direction soft red granite.

Other similar small tin lodes occur in various parts of the Cope's Creek district.

On the Bow-yard gully, between Tiengha and the Grove Station, Messrs. Canning and Hutton have discovered a tin lode which differs from the lodes above described, in its having a northerly strike; but this may be only a local variation, as it occurs in a broad belt of euritic granite, in which it may have connection with other lodes.

Near the surface this lode is only a few inches thick, but it increases to a width of nearly 1 foot, of almost pure tin ore, at a depth of about 20 feet, where it is broken by a fault or slide. The ore has a loose granular texture, and sometimes occurs disseminated in grains through the encasing rock, which consists of a whitish eurite much broken by joints. For a few inches on either side of the lode the rock shows alternate vertical layers of quartz and felspar. This feature I have once before noticed in a tin-bearing reef which I discovered near Middle Creek.

Splendid samples of tin ore, in large crystals of ruby red, amber, and other colours, have been obtained from the surface soil at the

Pride of the Ranges and Pine Ridge Tin Mines, Lower Cope's Creek. Small nests of tin crystals have been found in the granite, but no defined lode. A few miles lower down Cope's Creek the granite changes into a whitish, very coarse-grained rock, traversed by large quartz reefs bearing about E.S.E.

The Elsmore Mine I have twice inspected. It has already been described by Mr. G. H. F. Ulrich, F.G.S., in his paper on *Recent Tin Ore Discoveries*, read before the Geological Society. As a more faithful description could not be given, I will quote here an extract from it:—"Perhaps the richest mining area as yet discovered is that of the Elsmore Company, situated about 12 miles east of the boundary of Inverell. It lies on the north-west side of the Macintyre River, and includes a granite range about 250 feet in height and nearly 2 miles in length, dipping on all sides (save that towards the river, beyond which the rock extends a considerable distance) beneath basalt. The granite is micaceous, and rendered porphyritic by crystals of white orthoclase, which frequently reach several inches in size; bluish-grey oligoclase is also, though sparingly, associated. It is traversed by quartz veins several inches to above a foot in thickness, which contain cassiterite in fine druses, seams, and solitary crystals. Portions of these veins are highly micaceous, and represent in fact the rock called "Greisen," characteristic of tin ore districts of Saxony and Bohemia. In the tin ore localities of the Beechworth Gold Field, Victoria, this rock occurs also in a similar manner, but the surrounding granite is there very fine grained, and highly felspathic (euritic), and rarely shows porphyritic texture. Of far greater importance, however, than these veins, are dykes of a softer kind of granite, which consists perhaps for 75 per cent. of its mass of small scaly greenish mica, and the remainder of quartz; felspar being but very rarely observable. Through these micaceous dykes cassiterite is not only well distributed in implanted crystals from the size of a pin's head to above that of a pea, but it occurs also in irregular veins of several inches thickness, and in nests and branches yielding lumps of nearly pure ore up to above 50 lbs. in weight. Part of the mass of one of these dykes forms a regular breccia of mica and imperfectly crystallized tin ore, cemented by hydrous oxide of iron.

"The actual number of such dykes traversing the granite range is not known as yet. I saw six of them, each several feet in thickness, but there can be no doubt that more will be found when the ground is more minutely prospected than has hitherto been the case.

"As far as could be seen in the small workings of the Elsmore Company, on several of the quartz veins and dykes the dip of the latter is rather steep, and the walls pretty well defined, but thickness irregular. Thin flat veins join in occasionally. The deepest shaft sunk in one of the quartz veins was about 60 feet; and the tin ore occurred in irregular thin veins, and often beautifully crystallized in drusy cavities. On examining the spoil heap round this shaft, I discovered lumps of ferruginous clayey substance, full of light green and yellow hexagonal prisms of beryl, associated with larger quartz crystals. I also observed beryl on crystallized cassiterite specimens, its fragile prisms, generally not thicker than a stout pin, and up to an inch in length, interlaced between the tin ore crystals. Of other minerals, I found in the stuff excavated from one of the dykes frequently patches of arsenical pyrites, and, more rarely, grains of copper pyrites, the former generally containing imbedded crystals of tin ore. From another part of the ground the manager preserved a large piece of fine rock crystal, which also enclosed small crystals of the ore.

"Wolfram has been found at several places forming nests in the granite, but not in association with cassiterite. Touching the latter itself, it is mostly of a pitch black colour, occasionally translucent brown and hyacinth-red, and from some places greenish, with a very pretty play of rays of red and yellow colour through it. Its crystalline form is rather simple as regards pyramidal planes; the prism is generally, however, highly modified. Twins like those from the Schlaggenwald mines are very abundant, and crystals perfectly developed all round, both twins and simple ones—the latter with twelve-sided prism and one pyramid—are not rare amongst the ore washed from the drift."

A crushing of over 100 tons of the micaceous rock gave almost one per cent. of ore; and from one of the quartz reefs, at a depth of about 80 feet, the stone yielded from about 3 to 10 per cent. of tin ore. As there are several similar quartz reefs on the ground, the future prospects of this mine may be considered very encouraging.

The rock formation at the Newstead Tin Mine, which lies about 3 miles to the east, is of the same character as that of Elsmore. It consists of a porphyritic granite, traversed by tin-bearing quartz reefs and irregular beds and isolated patches of micaceous rock.

One of these small patches which I particularly noticed was about 18 inches in diameter, and thickly studded with crystals of tin ore. One of the quartz reefs has been opened, and contains besides tin ore, iron and copper pyrites, green carbonate of copper, and wolfram. The reef is from 2 to 3 feet thick, and bears about E.N.E. Very large crystals of quartz occur in it, some of which have crystals of tin enclosed within them, similar to those previously described from the Albion Mine.

The above remarks will also apply to the granite of the adjoining Karaula Tin Mine. All the granites of the district contain abundance of black tourmaline (schorl), which chiefly occurs in radiating masses, together with beautiful crystals, often of large size, of smoky quartz, in dense cavities in the rock.

In cutting a race through the granite at the Sydney Tin Mine, Middle Creek, "a blow" or sudden expansion of quartz was exposed. In the quartz I found abundance of fluor spar in imperfect octahedral crystals of an amethystine or greenish colour. With this were also galena, iron pyrites, and molybdenite. One small specimen I obtained with all these minerals together.

A reef of chalcedonic quartz occurs a short distance north of the Sydney Tin Mine, Middle Creek.

Small crystals of quartz in the form of a double hexahedral pyramid are of frequent occurrence in the granites.

Greenstone Trap.

The greenstone trap, already described in my former Report, occurs near Middleton, and extends for some distance on either side of Cope's Creek, as far up as Captain Swinton's Station. It may be traced in places between the granite and basalt boundaries, from the Macintyre

River across to Lower Cope's Creek. It also extends from Newstead Creek, for some distance to the westward, covered up occasionally by the older tertiary drift and basaltic trap. At Captain Swinton's, broad dykes of eurite traverse it (See figs. 7 *e* and 2 *h*), and in some localities quartz reefs are frequent. One quartz reef near the Pine Ridge Mine, Cope's Creek, contains copper ore, the green and blue carbonates, though not in payable quantity. The reef varies from 4 inches to 4 feet thick, and dips about S. 17° W. @ 80°. It is not improbable that some of the quartz reefs may be payably auriferous.

The foregoing observations show that the geology of the district of Inverell reveals many interesting facts connected with the physical history of this Country.

It has been remarked that the tin-bearing granites are of at least two periods of emission, and also that they are of later formation than the greenstone. (See fig. 7.) As to the age of the granites, the section at Newstead (fig. 6) indicates that they are newer than carboniferous, and this agrees with the opinion which the Rev. Mr. Clarke and other geologists have expressed on the subject. In his report (1853) the Rev. Mr. Clarke pointed out the resemblance which the New England granites bore to those which he had seen in the European Alps and in Devon and Cornwall. And again, in his Anniversary Address (1872) to the Royal Society of New South Wales, he states that "geologists at Home have settled it that the stanniferous granites are palæozoic, pre-permian, and post-silurian." Mr. David Forbes said, at the Geological Society's meeting in December, 1871, that he had received specimens of the granite from the New South Wales tin region, in the year 1859, and that he found them to be "perfectly identical with the stanniferous granites of Cornwall, Spain, Portugal, Bolivia, Peru, and Malacca."

Mr. G. H. F. Ulrich, F.G.S., has shown also that the micaceous veins at the Elsmore Tin Mine represent the rock characteristics of the tin ore districts of Saxony and Bohemia.

Sir Chas. Lyell (*Elemy. Geology*, p. 769), after referring to Sir H. de la Beche's Report on the Geology of Cornwall, states "that the most ancient Cornish lodes are younger than the coal measures of that part of England, although probably they are not newer than the beginning of the permian period." And speaking of the modes of occurrence and "faulting" of the tin lodes, the same author mentions (p. 761) that "it is commonly said in Cornwall that there are eight distinct systems of veins, which can in like manner be referred to as many successive movements or fractures. Both the tin and copper veins in Cornwall run nearly east and west. Many lodes in Cornwall and elsewhere are extremely variable in size, being one or two inches in one part and then 8 or 10 feet in another, at the distance of a few fathoms, and then narrowing again as before. Such alternate swelling and contraction is so often characteristic as to require explanation. "The walls of fissures in general," observes Sir Hy. de la Beche, "are rarely perfect planes throughout their entire course, nor could we well expect them to be so, since they commonly pass through rocks of unequal hardness and different mineral composition. If, therefore, the opposite sides of such irregular fissures slide upon each other, that is to say, if there be a fault, as in the case of so many mineral veins, the parallelism of the opposite walls is at once entirely destroyed." These observations, you will notice, almost equally apply to the tin lodes of New England. For I have already described several distinct systems of veins, joints, and faults, which prevail here; the general direction of our tin lodes is E.N.E.; and the alternate swelling and contraction of the lodes is observable in the Bolitho, and Canning and Hutton's tin lodes, and is, in fact, characteristic of nearly all the tin and other veins in the granite of this district.

The granites are said to attain an elevation of nearly 5,000 feet above the sea; and the fact that they are of upper carboniferous age is one of importance in considering the orographical features of the Cordillera.

The carboniferous formation of the district has been described as presenting a series of beds precisely similar in character to those of the Hunter and Hawkesbury, on the eastern side of the Dividing Range.

The identification of the older tertiary drifts, clays, and ironstones as of lower miocene age rests on their perfect lithological resemblance to certain leaf beds of that age in Victoria, and also on the fossil leaves and plant stems which have been found in the ironstones between Newstead and Elsmore. Similar leaf beds have been described by the Rev. Mr. Clarke (in his *Remarks on the Sedimentary Formations of N.S.W.*) as occurring in various parts of the Colony, and in one place at an elevation of 4,000 feet above the sea. He supposed them to be miocene, and observed that on comparing the living leaves with the impressions in the deposits mentioned he could see no specific identity. The impressions of leaves on the rocks near Newstead seem undistinguishable from those found in the above-mentioned leaf beds in Victoria. On the Geological Survey maps of Bacchus Marsh, Victoria, they have been described by Professor M'Coy as follows:—"The fossil plants of the ironstones are strikingly distinguished from the pliocene tertiary leaf beds of the Daylesford and other older gold-drift deposits, by the total absence of myrtaceous plants which so strongly mark the recent forest foliage of Victoria. I have no doubt the fossil leaves from this locality indicate a lower miocene or upper eocene tertiary flora, in which lauraceous plants form a remarkable feature. All the species seem new, but leaves of Laurus, Cinnamomum, Daphnogene, and possibly Acer, are scarcely to be distinguished from species referred to those genera in the leaf beds (of the geological age mentioned) of Rott, near Bonn, and Oeningen (specially the Cinnamomum polymorphum, Heer)."

These plant deposits therefore indicate the physical geography of this part of Australia to have been different in the miocene period from that which obtained in latter tertiary times.

In Victoria there interposes between these plant beds and the pliocene basalts a thickness of several hundred feet of marine fossiliferous strata. These are absent in the district I now describe; and the basalt, which is the next formation met with, is seen to have filled ancient valleys (see fig. 2, *d*), the erosion of which, since the deposition of the miocene plant beds, marks a lengthened period of even greater duration than that which succeeded the basalt eruption to the present time.

Of the latter period its duration may be imagined by contemplating the time required for rain and river action combined to erode a valley several hundred feet deep through basalt, granite, and other rocks, as has been the case with the present Macintyre Valley.

These hills also bear testimony of the vast denudation the land has suffered during the pleistocene period. Besides the scattered outliers of basalt now capping the hills, the granite rocks afford abundant evidence of long continued erosion and decay by atmospheric influences.

The accompanying sketch (fig. 10) is one that I took of some remarkable granite rocks near Middleton, Cope's Creek.

The three rocks are together about 20 feet high. A branch of a tree has grown against the upper one, and appears now to support it from falling. The middle rock (calculated from its measurements) weighs 45 tons. The granite is of the ordinary ternary kind—quartz, felspar, and black mica; the felspar predominating, and the mica scarce. The western side (that on the left hand in the sketch) of that rock has a smoother surface than that facing to the east.

The sketches figs. 11, 12, 13, kindly taken at my request by Mr. Licensed Surveyor R. L. Murray, are of some rocks at the Elsmore Tin Mine, and near Cope's Creek. These also show in a remarkable manner weathering action on granite. Other similar instances are frequent throughout the tin-bearing country.

They now well serve to indicate the depth to which the surrounding rock, of which they once formed part, has been removed by denudation; just as the little pillars sometimes left in earth-cuttings by the navvies mark the depth to which the surrounding earth has been excavated.

In addition to the above-mentioned interesting physical features, the facts which the geology of the District of Inverell reveals have an important economic bearing.

In the carboniferous formation the finding of "coal in some abundance" has been predicted. Coal would no doubt be a valuable acquisition to this tin-mining district.

I have mentioned the discovery of a number of diamonds on the Borah Creek, where I anticipate many more will be found; and their occurrence in various other parts of the district proves that they are pretty widely distributed. There seems but little doubt that they have been derived from the older tertiary gravels; and this is in agreement with the observations of the late Professor Thomson and Mr. Norman Taylor on the Cudgegong Diamond Field. For the fullest information, however, on this subject, I would refer to the valuable remarks of the Rev. Mr. Clarke on the History of the Diamond in Australia and Foreign Countries, in his Anniversary Address to the Royal Society of New South Wales, May, 1872.

As regards the tin-bearing resources of this district, its future prospects are I consider very auspicious. The amount of tin ore raised during 1872 was about 800 tons. This yield would doubtless have been larger but for the wet spring season, and the unsteady working of the mines consequent on the excitement which the tin discoveries created.

Should the weather be favourable, the yield during the present year will I believe exceed three times that of the last; for, in the neighbourhood of Tiengha alone, I have been informed that 50 tons of ore have lately been obtained in one week. It is, however, impossible to foretell accurately what quantity of tin ore may be annually raised.

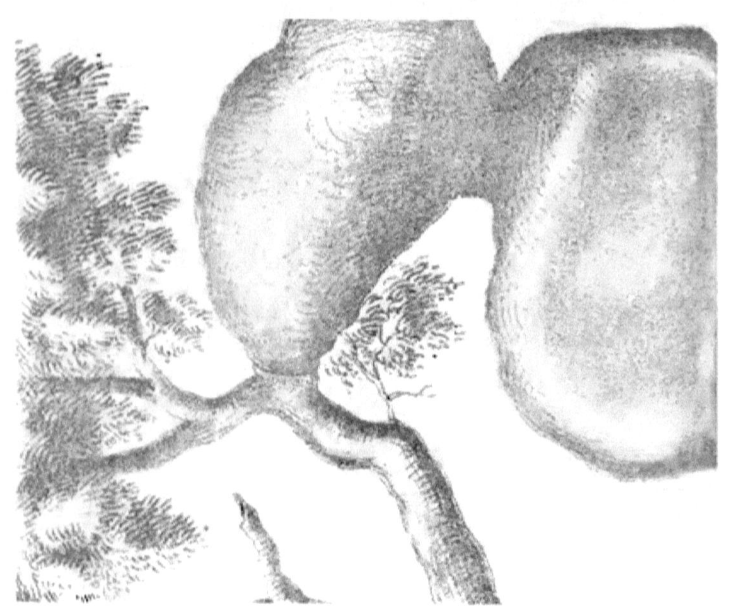

DRAWN BY C S WILKINSON, G.S.　　PRINTED AT THE SURVEYOR GENERALS OFFICE, SYDNEY N

PRINTED AT THE SURVEYOR GENERAL'S OFFICE, SYDNEY NSW

g weathering of Granite,

ear Copes Creek

Remarkable Rocks _ shewn
Sutherland's Water,

g weathering of Granite.
near Cope's Creek

The valuable deposits of stream tin which both the recent and older tertiary formations contain will no doubt take many years to work out; and those deposits which, at the high price of labour and with the present imperfect appliances for extracting the tin ore, will not now pay to work, will doubtless be developed in the future.

The tin lodes as yet discovered have been described as having characters identical with those of the lodes of Cornwall. This fact alone should encourage the enterprise of prospectors, and must also give assurance of the permanency and importance of tin-mining in the district of Inverell.

And seeing that the district I have endeavoured to describe (that within a radius of 25 miles from Inverell) forms but a small portion of the stanniferous country of New South Wales and Queensland, we may readily conceive the future magnitude which the tin-mining industry in these Colonies is destined to attain: indeed, the annual yield from the Australian tin mines is, even at the present time, about equal to half that of all the old tin-mining Countries combined.

During my examination of this district, I collected characteristic specimens of all the rocks and minerals (especially of the tin ores) mentioned in this Report, and I have arranged them in the Geological Museum of the Crown Lands Office. Some of them it would be well to have analyzed.

In such an important mining Country as New South Wales is now becoming, a Government laboratory would doubtless be of great advantage; by contributing knowledge of the value of our mineral resources, it would materially promote the mining interests of the Colony.

COPPER.

Considering the number and extent of the copper lodes in this Colony, it is surprising that so small a quantity of ore has been raised and treated. It is, however, gratifying to find that the attention of capitalists is now being attracted to this branch of mining, and it is to be hoped that not only will a great number of lodes be developed, but that the facilities for selling or treating the ores will be largely increased. If a demand for copper ore, at a fair price, were created, many of our miners who do not possess the means of erecting smelting works of their own would turn their attention to copper-mining, and would find profitable employment. In many of our lodes other metals are found with the copper, and if the whole of these could be separated and saved by an inexpensive process, the profits derived from our copper mines would be largely augmented. It is quite possible that many of our copper lodes in remote districts, though known to contain rich ore, will remain unworked for some time to come, owing to the cost of bringing the ore to market, unless the necessary capital can be procured to erect smelting works in the locality, but there are large tracts of cupriferous country through which the projected lines of railway will in the course of a few years be carried, which, with the facilities afforded thereby, will give profitable employment to a very large number of miners, and will raise the yield of copper to an extent commensurate with the value of our deposits.

COPPER, the produce of the Colony of New South Wales.

	Copper.		Copper Ore.		
Year.	Quantity.	Value.	Quantity.	Value.	Total value.
	Tons.	£	Tons.	£	£
1858	......	....	58	1,400	1,400
1859	......	....	150	2,250	2,250
1860	......	....	43	1,535	1,535
1861	......	....	144	3,390	3,390
1862	......	....	2,200	12,000	12,000
1863	125	12,500	....	....	12,500
1864	......	....	2,100	22,100	22,100
1865	295	29,491	1,618	7,854	37,345
1866	304	23,390	947	4,745	28,135
1867	296	19,866	2,590	15,450	35,316
1868	315	21,420	3,151	12,780	34,200
1869	324	21,446	1,437	5,400	26,846
1870	297	20,060	84	336	20,396
1871	665	47,231	2½	44	47,275
1872	419	36,770	1,466	17,873	54,643
1873	150	14,500	5,877	142,126	156,626
1874	3,638	311,519	....	....	311,519
	6,828	£558,193	21,897½	£240,283	£807,476

The quantity of copper ore *raised* during 1874 could not be ascertained. That given above is the quantity of copper *exported*.

The following extracts from reports furnished by Mining Registrars and others will convey some idea of the extent to which copper-mining has been carried, though unfortunately the reports are very far from complete, and no return has been made from some important mines.

The Belara Copper Mine, 20 miles from Gulgong, county Bligh; 150 tons at grass; no analysis yet made; lode 2 feet thick; strike N. and S.; dip 2 in 6.

The Frogmore Copper Company, at Frog's Hole, in the parish of Bula; six men employed; 70 tons of ore raised; no analysis yet made; the average thickness of lode at 75 feet is 3 ft. 6 in., consisting of carbonates and sulphurets; strike of lode north-west; no analysis yet made; ore is estimated to contain 12 per cent.; the lode is regarded as very promising. There is another copper mine in the same locality, of which no particulars are to hand.

Peelwood Copper Mine, at Peelwood, 10 miles from Tuena; 100 men employed; thickness of lode 4 feet; strike N. and S.; dip 15° easterly; present depth 384 feet; the foot-wall consists of schist, the hanging-wall of soft slate; some 1,900 tons have been raised, of which 1,100 tons have been smelted, for which purpose three furnaces and a calciner are used. The results obtained by analyses are from 7 per cent. to 26 per cent. The average yield obtained is 12½ per cent. Silver and lead are also found, but are not worked at present.

Near Orange there are several copper lodes. The Great Western Copper Company are working the following lodes at Icely, viz.:—The Williams Lode, which is 2½ feet thick at the surface, 5 feet at 25 fathoms, and 3 feet at 45 fathoms; the strike is 30° west of north; the dip is easterly, 1 in 4; 4,000 tons of ore raised, yield 560 tons of copper. Upon this lode they employ twelve men. The Icely lode, which has been proved to a depth of 25 fathoms, average thickness 1 ft. 6 in.; strike 30° west of north; dip 1 in 6, easterly; 1,000 tons of ore raised, yield 100 tons of copper; matrix of ore, clorite, and copper pyrites. Upon this lode they employ thirteen men. There are several other lodes, of which they have worked four; the strike and dip of these are the same as above; from these they have raised 100 tons of ore, which has given 12 tons of copper. These lodes are enclosed by blue slate. Upon these lodes they employ six men. The same Company are working eight lodes at Lewis Ponds Creek; average thickness 3 feet; strike 20 west of north; dip 5 in 24, easterly; proved to a depth of 28 fathoms; 4,000 tons of ore raised, yield 640 tons of copper. Upon these they employ ten men; matrix of lodes clorite; lodes enclosed in clay-slate; ore smelted by fire, and brought on by heat and air in reverberatory furnaces.

The Ophir Company is working a lode at Brown's Creek, average thickness 2 feet; strikes 30° west of north, dip 1 in 6 easterly, proved to a depth of 25 fathoms, matrix quartz. 200 tons of ore raised, yield 30 tons copper; number of men employed 20. The ore contains gold, also oxide, carbonate, and sulphuret of copper.

The Carrangara Company is working a lode at Byng. At a depth of 5 fathoms the lode is 3 inches thick, at 10 fathoms 6 inches, at 30 fathoms 1 foot, at 40 fathoms 1 foot 6 inches; the strike of the lode is 15° west of north; the dip 60° west. 5,600 tons of ore have been raised, of which 2,240 tons have been treated, yielding 784 tons of copper. Analysis shows the ore to contain quartz, iron, sulphur, copper and gold. The mode of treatment is as follows:—The ore is calcined in the open air, then fused in a reverberatory furnace, skimmed, tapped, roasted, and then refined,

The Wiseman's Creek Company is working a copper lode at Wiseman's Creek, which has been proved to a depth of 120 feet. The lode at the surface is about 20 feet thick, and at 120 feet it is 18 feet thick; the strike of the lode is north by west, and the dip is slightly to the west; 700 tons have been raised, of which 350 tons have been sold to a smelting Company, at an average of £7 per ton. The result of analyses of ore taken indiscriminately from the pile, without dressing, is 9 to 17·75 % of copper, the same of lead, and traces of gold and silver; number of men employed is 14.

The South Wiseman's Company have proved the same lode, about one mile south of the above Company, to about the same depth. The result of analyses is 9 to 20 % of copper, same of lead, and traces of gold and silver. They have raised 200 tons of ore, which they have sold to a smelting Company, at £6 10s. per ton.

The Armstrong Company are working a copper lode in the parish of Yetholme. The lode is 2 feet thick at the surface, 3 feet at a depth of 30 feet, and 4 feet 6 inches at 45 and 50 feet. The strike of the lode is east and west, the dip slightly inclining to the north. Between 60 and 70 tons of ore have been raised, but the ore has not yet been treated, as it is the intention of the Company to erect smelting works on a new principle, which was patented in America, and more recently in this Colony. The result of analysis proves that the ore contains from 15 to 36 per cent. of copper, besides a large percentage of gold and silver. The ore has changed rapidly from carbonate to red oxide, then to peacock and grey ore, containing a high percentage of copper, and they are now on black ore, containing 20 %. Number of men employed eight.

The Apsley Copper Company are working a copper lode in the parish of Apsley. The lode has been proved to a depth of 100 feet. The lode is 5 feet thick to a depth of 50 feet, and from that depth to 100 feet it is 3 feet; the strike of the lode is north and south, and the dip is easterly. The result of analyses gives 4 % to 18½ % of copper; they are raising 150 tons monthly. The ore is sold to the Esk Smelting Company, at the rate of £4 5s. per ton, for 10 %, rising 13s. per unit, and 4d. for every £2 over £85. Smelting works have been erected on the old plan, at a cost of from £700 to £800, on purchased land, adjoining the mine; number of men employed eighteen.

In the Wellington District a considerable area of copper land has been taken up recently, but in most cases work has not yet been commenced. At Gordon Brook, in the Clarence District, a Company has recently started, and has raised 12 tons of ore, but none of the ore has been treated. Some very promising lodes have been opened in the Bingera District, but there are no facilities for bringing the ore to market.

SILVER.

It is clear that the returns of the quantity and value of the silver ore raised and of the silver extracted are very incomplete, indeed there is no record of the quantity produced between 1865 and 1869. Inquiries will be made with a view to supply the defects in the returns, but it is not possible to complete such inquiries in time to embody the results in the table, which will therefore be accepted as representing only some portion of the silver produced.

SILVER, the produce of the Colony of New South Wales.

Year.	Quantity.		Value.
			£
1862	266 tons (ore) say		5,320
1863	28 " "		1,080
1864	13 " "		130
1865	736 ozs. silver.		184
1866	Nil.		
1867	Nil.		
1868	Nil.		
	ozs.	dwts.	
1869	753		199
1870	13,868	6	3,801
1871	71,311	18	18,681
1872	49,514	17	12,663
1873	66,997	10	16,278
1874	78,027	0	18,880
	307 tons ore + 281,238 ozs. 11 dwts. silver.		£77,216

ANTIMONY.

But little attention appears to have been paid to the extraction of this mineral, though it has been discovered in various parts of the Colony in such quantities as should, under favourable circumstances, render the working of it a profitable employment. As the mineral resources of the Colony come to be better known and appreciated, there is no doubt antimony will be raised in larger quantities, and will form an important item in our returns.

Antimony, the produce of the Colony of New South Wales.

Year.	Quantity.		Value.
	tons	cwt.	£
1871	31	0	560
1872	0	13	5
1873	27	12	210
1874	12	15	122
	72	0	£897

IRON.

Every exertion has been made to obtain a complete and reliable statement of the quantity and value of the iron ore raised, and of the iron smelted, but the attempt has not been successful. So far as can be ascertained, some 3,600 tons of calcined ore from the Fitzroy Mine gave 2,394 tons of pig iron, which sold on an average for £6 per ton. The iron ore is said to be a very fine quality, but somewhat difficult to smelt; it yields after calcining about 66 per cent. of iron. The Lithgow Valley Iron Works are progressing very satisfactorily, and will probably be complete in the course of a few months. At Brereton's Mine, Berrima, some 1,000 tons of ore were raised during the year 1874.

The following paper on the Iron Ore and Coal Deposits at Wallerawang, New South Wales, by Professor Liversidge, University of Sydney, was read before the Royal Society, 9th December, 1874:—

Many are probably well aware that there are large deposits of iron ores and extensive beds of coal in the neighbourhood of Wallerawang; but comparatively few, perhaps, are in possession of any very definite information concerning them. I therefore beg to lay before you the substance of some notes, taken during a brief visit which I made in the early part of August last, and the results of my subsequent examination of the samples of the ores and coals which I then collected.

I much regret that I cannot afford any general and comprehensive account of the geology of the district; and that it is only in my power to speak definitely upon the actual deposits of iron ore, coal, and the closely associated limestone. For, owing to an unfortunate accident which I met with to my foot, within a few days after my arrival at Wallerawang, I was entirely prevented from making any detailed investigation of the various strata and deposits other than of those which follow.

I very much regret too that I had to relinquish the idea of working out the geological section of the district; but as Mr. Chas. Wilkinson, the Geological Surveyor, recently appointed by the Government, is now engaged making a survey of the district, we shall probably soon be in possession of a complete report upon the whole of the coal measures and iron deposits of this area, and in a much fuller and more detailed form than I could possibly have hoped to have worked it out in the time at my disposal, consisting as it does merely of short intervals of leisure.

Wallerawang is distant from Sydney some 105 miles, on the western line of railway.

The township and station of that name is situated on a "drift" composed of pebbles disseminated through a soft argillaceous cement or clay. The enclosed pebbles consist principally of rolled fragments of quartz, jasper, flinty-slate, argillaceous sandstone, and other substances. On the whole, this drift bears a very close resemblance to the diamond-bearing drift of Bingera and other places; and, like the diamond drift, it contains nodules of conglomerate, composed of rounded and sub-angular fragments of white and coloured quartz, and various other minerals, agglutinated together into a compact mass by a ferruginous cement.

It also contains a small quantity of gold, but apparently not in sufficient quantity to pay for working at the present time.

This drift can be traced for some distance—as far, I am informed, as Bathurst. Good sections of it are seen at the Wallerawang Railway Station, and along the Mudgee and other roads near the town, where several small cuttings show its structure very well.

The deposits of iron ore at present opened out are situated some six miles from Wallerawang, and near the junction of the coal measures with the Upper Silurian or Devonian beds, which there crop out to the surface. These deposits contain two varieties of iron ore, viz.—magnetite or the magnetic oxide of iron, and brown hematite or goethite—the hydrated oxide; then in addition to these there are deposits of the so-called "clay band," which are interstratified with the coal measures. These clay bands are not what are usually known as clay iron ores in England. They are brown hematites, var. limonite, while the English clay iron ores are impure carbonates of iron, which seldom contains much more than 30 per cent. metallic iron, against some 50 per cent. contained by the hematites.

A highly ferruginous variety of garnet accompanies the veins of magnetite; this garnet is very rich in iron, and it will probably be found advantageous to smelt it with the other ores, not only on account of the large percentage of metal which it contains, but also on account of the increase fluidity which it would impart to the slag.

Iron Ore Deposits.

1. *Magnetite.*—The vein of magnetic ore runs apparently N.E. by S.W. This can only be stated approximately, for, owing to the action exercised by it on the needle, the compass was found to be perfectly useless in the vicinity of the lode,

The ore is scattered over the ground in blocks and nodules along its outcrop; but at a little depth it is in a solid and compact body, merely broken across here and there into large masses by joints and fissures.

In one part the vein has a width of thirteen (13) feet; but at another spot, where a trench was cut across, it was there found to be not less than 24 feet in width.

Two shafts have been sunk on this vein—one at a depth of 10 and the other to a depth of 23 feet. At these depths the quality of the ore is about the same as that at the surface.

Certain portions of the vein are evidently richer than others.

At present the average yield of metallic iron from the vein, as a whole, is not rich for a magnetite, which, when perfectly pure, contains 72·41 per cent. of iron, and under ordinary circumstances about 70 per cent., whereas the Wallerawang vein yields only 40·89 per cent. (*See analysis appended.*) Although this is a poor magnetite, it must not be regarded as a poor ore of iron.

This average was obtained by taking samples from different parts, across the whole width of the trench cut across the vein, and then crushing them all up together. As I have before mentioned, picked portions yield a much larger percentage.

On the whole, taking all the circumstances into consideration, we may come to the conclusion that the true capabilities of the deposit of magnetite have not yet been fully tested or proved.

The vein stuff or gangue accompanying the magnetic iron ore is silicious. In some parts of the lode this appears to be replaced by the ferruginous garnet rock.

On analysis this ore yielded the following results :—

Silica and insoluble matter... ...	18·70 per cent.
Metallic iron	40·89 „
Phosphorus	Traces.
Sulphur	Traces.

Both the phosphorus and the sulphur are present in such minute quantities that the ore may be regarded as virtually free from them; and these are the only really deleterious substances present; for although there is too large a quantity of silica and gangue present in this superficial portion of the vein to permit of malleable iron being made from it by a direct process, it is extremely well adapted for reduction in the blast furnace.

2. *Garnet.*—The garnet occurs both crystallized, in the form of the rhombic dodecahedron, and in the massive state. The crystals are, as is usually the case very uniform in size; they are nearly all of them either about ¼ or ⅛ of an inch in diameter.

The faces of the crystals are smooth, free from pits and irregularities, and bounded by sharp and well-defined edges. The colour is brown without any red shade.

Portions of the massive garnet and aggregations of crystals are hard and compact, whilst in other parts they are more or less disintegrated and friable.

The average percentage of metallic iron is 21·05—an amount not much less than that contained by many commonly smelted ores.

3. *Brown Hematite.*—The general direction of the outcrops of this deposit is not so regular as that of the magnetic lode, and it will probably be found that other veins run into it, but for a large portion of its course it runs approximately N.E. by S.W.

Although the back of the lode does not absolutely come to the surface along its entire course, yet there is a great probability of all the different outcrops being connected beneath the surface—in which case the total length of the deposit, as far as it has been at present traced, cannot be much less than one mile.

Along this line of outcrop the ore is seen scattered all over the surface in great blocks and nodules, either completely exposed or but partially embedded, and over a width of from 12 to 18 feet in parts, to as much as even 50 or 60 feet in another.

The thickness of the deposit below the surface has not yet been fully ascertained; but a shaft has been put down in the lode itself to a depth of 43 feet, and at the bottom a level was driven which proved it to be of the same quality through a thickness of 18 or 20 feet. There are no decided appearances of the boundaries of the deposit having been cut, except on the N.E. side, so that it may eventually prove to exceed the above-mentioned thickness.

As far as can be seen from the outcrops and other indications, the deposit has nearly every appearance of being a true vein or lode, in which case of course its depth may be regarded as practically unlimited, for there is nothing on record which shows that the bottom of any true vein has ever been reached in mining operations. Veins have often been regarded as worked out altogether when reduced perhaps to a mere thread, but, on continuing to sink, such veins have always been found to open out again. One cannot of course say that this will always be found to be the case, but it always has been the case, when sufficient perseverance has been used in following the traces of the vein downwards.

However, although there are strong indications of this deposit of brown hematite being a true lode, it has not yet been conclusively proved to be one; subsequent workings may show it to be an irregular deposit such as those of the Forest of Dean and other places in England.

As the shaft already put down is sunk through the deposit itself, the stuff raised and heaped up at the shaft mouth affords a very fair sample of the average quality of the ore.

The ore is composed of nodules of mammillated and botryoidal goethite, possessing a fibrous structure something like that of wood, mixed with massive and friable brown hematite, together with a little pipeclay. On descending the shaft, its walls, on all sides and from top to bottom, are seen to be composed of the same ore, only, of course, from not having been disturbed it is much more compact. Near and at the bottom, a "horse" of pipeclay comes in, but one of no great size.

A sample of the ore taken from the heap at the shaft's mouth, just as it was raised, and without having undergone any dressing process, yielded on analysis the following results:—

Analysis of loose brown Hematite.

Water, hygroscopic	2·74
„ combined	9·70
Silica and insoluble matter	25·33
*Sesquioxide of iron	54·23
Phosphrus	·27
Sulphur...	·12
†Undetermined	7·61
	100·00

* Equivalent to 37·84 per cent. metallic iron.

† Consisting principally of manganese, alumina, lime, and magnesia.

If for any special purpose a richer or purer ore be required, the nodules could be readily separated from the loose ore before it is sent to the surface. As shown by the following analysis, these nodules are much richer, and contain on an average 51·52 per cent. metallic iron.

Analysis of Goethite and massive Hematite.

Water, hygroscopic	1·28
" combined	12·04
Silica and insoluble matter	12·19
*Sesquioxide of iron	73·60
Phosphorus	·12
Sulphur	·06
†Undetermined	·71
	100·00

As will be seen from the above analysis, this brown hematite contains but very small proportions of phosphorus and sulphur, and far less than the well known Northamptonshire brown iron ore.

Probably it will be found feasible and economical to work much of this deposit as a quarry in stopes, on account of its great width, and because it is favourably situated on a hill side. By this method expensive shafts and galleries can be dispensed with.

4. *Clay bands.*—As I have already mentioned, the English clay iron ores, or "clay bands" as they are generally termed, are impure carbonates of iron, containing a large quantity of argillaceous or clayey matter. In Scotland an ore of this kind occurs, containing a large quantity of carbonaceous matter as well, and is known as the "black band," and like the English "clay bands" it is found interstratified with the coal measures. The percentage of carbonaceous substance varies from 10 to 15 per cent., and in some cases rises to even 30 per cent. The presence of this is a great advantage, for they contain sufficient to effect their roasting, previous to reduction, without the addition of any extra fuel.

This so called "clay band" appears to be more of a brown hematite, of the kind kuown as limonite.

I was enabled to examine four seams of this ore. They are interstratified with the coal measures, and, in common with them, at this part they are approximately horizontal, having only a slight dip of not more than about 2° to the N.E.; their outcrop are seen jutting out in the gullies and creeks on both the E. and W. sides of the Dividing Range.

The lower of the four seams is perhaps the least pure and valuable of them all, but the other three are of very good quality and of great value, as shown by the analysis.

* Equivalent to 51·52 per cent. metallic iron.
† Consisting principally of manganese, alumina, lime, and magnesia.

Analysis of clay band iron ores.—Clay band No. 1.

Water, hygroscopic	1·28
" combined	3·54
Silica and insoluble matter	4·60
*Sesquioxide of iron	80·00
Phosphorus	·19
Sulphur	·11
†Undetermined	9·98
	100·00

Clay band No. 2 contains 53·31 per cent. metallic iron.

Clay band No. 3 contains 49·28 per cent. metallic iron.

It is highly probable, from the unusual richness of clay band No. 1, and from the small quantity of combined water which it contains, that it has been subjected to a bush fire, and answers, therefore, more or less, to roasted ore. A sample of this ore, taken from an unexposed portion of the seam, would most likely yield about 50 per cent. metallic iron, in place of 56 per cent.

As is usually the case with such deposits, the thickness varies somewhat; in some places they are from 8 to 9 inches thick, and then a little farther on they widen out to a thickness of even 18 inches. The average thickness of the two lower seams may be taken at about 10 inches each, and the upper seam (No. 3) at 11 or 12. These dimensions are estimated from the outcrops of the seams, and are of course only approximate, for as they have not yet been opened out or cut across in any way, no clean sections are exposed, and consequently no minute measurements could be taken.

It is at once apparent from the analysis that they are all three richer than the ordinary English clay band ore, and that the amounts of phosphorus and sulphur, for all ordinary purposes, are unimportant.

Coal.

The coal measures in this district contain several very valuable and thick beds of coal. The three principal ones which I had the opportunity of closely examining have respectively the following thicknesses of coal irrespective of any partings:—

The lowest seam, which I will call No. 1, has a thickness of 17 feet 6 inches; the next seam, No. 2, is 6 feet 6 inches; and the one above this, No. 3, is 4 feet 6 inches.

There are other seams present, but as they are thinner they are of minor importance, and in the face of the above thick seams they are not likely to be touched for some years.

Seam No. 1—The outcrop of this bed is seen in the banks on a creek known by the name of Coal Creek, on the western side of the Dividing Range.

A trial shaft sunk through it has proved it to be 17 feet 6 inches in thickness, divided by a parting of fire-clay some 8 inches thick.

The parting of fire-clay shows the numerous remains and impressions of coal measure plants—principally thin rootlets in this case—embedded in the original soil in which they grew. At this period of the history of the coal bed there must have been a change in the conditions throughout the area over which this parting extends; the

* Equivalent to 56 per cent. metallic iron.

† Consisting principally of manganese, alumina, lime, and magnesia.

circumstances had become unsuitable for the continuance of the growth of luxuriant vegetation which previously covered it. This unfavourable change may have been brought about by a variety of causes. It was most probably due to a gradual depression of the area beneath the surface of the water, which period of depression extended sufficiently long to allow of the deposition and accumulation of the eight inches of finely divided mud and silt which was the original form of the fire-clay. It is generally regarded that coal has been derived from the decay of terrestrial plants which flourished in marshy places, and that the majority of them consisted neither of true land nor of true aquatic plants, but of such as go to form the peat mosses, the mango and other swamps of the present day; hence a considerable depression of the area would be inimical to such growths. After this process of depression had gone on for a certain time, then the area was again slowly upheaved and the remaining eight or nine feet of coal was accumulated.

The quality of the coal is very good: it is hard and compact, and would therefore be well adapted for certain metallurgical processes, especially for use in the blast furnace, where it would have to sustain a great weight, and under circumstances where ordinary tender bituminous coal would have to be previously coked.

It possesses a sp. gr. of 1·333.

Analysis of 17 ft. 6 in. seam of coal.

Moisture	...	1·51	
Volatile hydrocarbons	...	33·24	
Coke { Fixed carbon	...	55·74	} = 65·24 coke.
Coke { Ash, white	...	9·50	
		99·99	

It is very free from sulphur.

This bed, in common with the others, is nearly horizontal; it has, however, a dip of about 2° to the N.E.

Seam No. 2.—An outcrop of this bed is seen in Coal Gully, and an exploratory level has been driven into it to a distance of about 60 feet.

At the outcrop, where cut by the level, it is seen to be about 6 feet 10 inches in thickness, with a 2-inch parting of fire-clay, which, however is gradually pinched out as the level proceeds inwards, and finally disappears altogether on the face.

The roof is a hard and compact sandstone. Throughout their entire thickness the coal measures consist of alternate beds of fire-clay or shale—the original soil on which the coal vegetation grew—coal and sandstone succeeded by shale, then coal again, and so on. Occasionally the order may be slightly altered, but in the main the series is continued throughout in that way.

In quality the coal is almost identical with the former one, but as is shown by the analysis, it contains rather more combustible matter and less ash. Like the former, it is very free from sulphur.

Coal from Seam No. 2. (6 ft. 6 in.)

Moisture	...	1·95	
Volatile hydrocarbons	...	27·25	
Coke { Fixed carbon	...	61·86	} = 70·80 coke.
Coke { Ash, white	...	8·94	
		100·00	

It possesses a sp. gr. of 1·398.

Seam No. 3.—This seam has a thickness of 4 feet 9 inches, with a 3-inch parting, leaving 4 feet 6 inches of coal.

It is rather a brighter and more tender coal than the others, and will probably be found well adapted for household purposes.

It occurs at a height of about 76 feet above the seam No. 2, or 6 feet 6 in. bed, while that in turn is about 118 feet above the seam No. 1 or 17 feet 6 in bed.

Clay band No. 1 is situated some 12 feet above this No. 3 seam or 4 feet 6 in. bed, and the other two clay bands are a little higher still.

One of the seams of coal crops out on the Mudgee Road about $2\frac{1}{2}$ miles distant, but as I did not take the levels this requires confirmation. This seam is worked, apparently on no large scale, by levels driven in from the road side,

Limestone.

Limestone.—Between the iron ore deposits and the coal seam outcrops there is seen an outcrop of limestone abutting against Devonian or Upper Silurian slates. Both the slates and the limestone are here standing at a high angle. The limestone does not show the dip so distinctly as the slates, for the lines of bedding have been almost completely obliterated, but the dip appears to be about 75° to the eastward, and the strike nearly N. & S.

At the junction of the two the limestone has evidently undergone much disturbance and is much brecciated, and includes within it fragments of the slate. Some of the included slate contains small crystals of iron pyrites disseminated through it.

In colour the limestone is of a bluish-grey or slate-colour, much veined with white calcite. The slate-coloured portions break with a slight crystalline appearance, but the calcite veins show the rhombohedral cleavage of that mineral on a large scale.

Its extension can be traced for a long distance to the north.

Not far from the small quarry which has been opened out in this limestone on Brunt's Creek, perhaps a hundred yards or so to the right hand after crossing over the creek, there is the opening to a fissure in the limestone. On the surface, just level with the ground, there is a small somewhat circular opening, surrounded and overgrown by grass and bushes—so much overgrown that it is almost completely hidden. This is probably the entrance to a cave in the limestone, for they usually afford no more indication of their presence than such a grass-grown aperture. The opening is only wide enough to allow of a man lowering himself some four or five feet, but from that point a narrow fissure can be seen to descend for some depth. Diligent search might prove that there are other openings, and the cave to be of some extent—and there is of course a very fair chance of its being found to contain animal remains; so that I hope Mr. Winter, of Wallerawang, who pointed it out to me, will be induced to follow his discovery up. The narrowness of the opening is not at all unfavourable to the supposition that the cave may eventually prove to be an extensive one, for very few limestone caves present large and well-marked entrances.

Caves in limestone have usually had their origin in fissures, through which water flows or at one time flowed: the water at first slowly percolating through them, and then as the fissure gradually became larger and larger the volume of water likewise increased until the fissure became converted into a true underground river or watercourse. Even in cases where no water flows through them at the present day it can plainly be seen that such was the case once, as exemplified in the caves or "swallow-holes" of Yorkshire, the Katavothra of the Morea, South Australia, and many other parts.

These caves are eaten out of the limestone by the solvent power which water charged with carbonic acid possesses. Ordinary water free from carbonic acid would be quite incapable of dissolving out the limestone, but all natural waters contain more or less of that gas, derived by the rain from the atmosphere as it falls and from the decaying vegetable matter which it meets with in its passage through the soil.

All limestone caves usually retain more or less completely their original form of fissures, expanded, perhaps, in parts into vast caves and chambers of immense proportions, but again contracting a little further on into a mere crack or tunnel.

Comparatively large rivers are received by such caves, which then continue their course under ground, in some cases suddenly appearing to the light of day again, but in others making their way beneath the surface right out to sea. Certain of the South Australian creeks are thus discharged.

I do not refer to this subject of caves in limestone so much on account of the supposed one at Wallerawang—for it may quite likely be proved to be merely the beginning of one—as to draw attention to the occurrence of such apertures in limestone districts, in order that they may be properly investigated.

In conclusion, I think I may safely say that this portion of the district of Wallerawang seems to be destined to be one of the greatest and most flourishing portions of the Colony. Here, within a comparatively small circle of some four miles diameter, there are extensive and rich deposits of iron ores, coal, and abundance of limestone. At present nothing beyond exploratory work has been done with them; but as the Wallerawang Iron and Coal Company has taken up large selections of the lands for the purpose of erecting iron works, there is a prospect that in a short time an attempt may be made to utilize some of this great wealth.

The whole of the district along the western line, near to and beyond Hartley, is one of exceeding interest to the geologist from a purely scientific point of view, quite apart from the importance and actual intrinsic value of the various mineral deposits which it contains.

It is a source of great gratification to all who take any interest in these matters that, at last, the resources of this and other portions of New South Wales stand a fair chance of being thoroughly and properly examined, now that the first step towards having a geological survey of the country has been taken by the Government—a step which may be regarded as an earnest of something to follow on a more comprehensive and extended basis; for of course it is utterly impossible for any one geologist, however great his attainments, to make single-handed a finished survey of a country like this.

No one will deny that money spent upon such an object is spent in one of the best possible ways, whether it be purely for the extension of scientific knowledge or merely for the exploration and development of the mineral wealth of the Colony. Perhaps the truest wisdom is to keep both ends in view:—the extension of science would make but comparatively little progress without the aid of wealth, and wealth, at the present day, cannot be attained without calling in the aid of science—they are mutually dependent, and on that account we cannot afford to neglect either of them.

The exploration and development of the mineral wealth of a country should always be kept a long way in advance of the work of realizing and converting such stores into money.

When we consider the great repositories of iron ores which have been already examined in New South Wales, and that we hear of discoveries of others, perhaps equally extensive, there appears to be no reason why New South Wales, with proper care and management, should not very soon make not only all the iron required for its own consumption, but also supply other countries which are not so lavishly endowed.

GEM STONES.

Considering the variety and number of gem stones which have from time to time been found in various parts of this Colony, it is to be regretted that no returns can be furnished of the number and value of those found. Though they are known to exist in certain localities, no attempt is at present made to collect them. Mr. Maskelyne, the learned mineralogist in charge of the mineralogical collection at the British Museum, says:—"Doubtless hundreds of diamonds have been washed away to perdition in gold washings, for old Pliny's statement, 'that diamonds and gold occur together,' holds good down to our time in every great diamond, if not in every gold country." Under existing circumstances, it may perhaps be questioned whether the miners can make the search for gem stones exclusively a profitable employment, but it is highly probable that they would be found in many of the auriferous drifts now being worked if search were made for them, and such search, if successful, could scarcely fail to be profitable, because in the process of gold washing, nearly the whole of the work is done. For example, where the sluice is used, only in that part of the sluice in which the gold would be likely to be deposited, would search need to be made, because the specific gravity of nearly all the valuable gems is greater than quartz, gravel, &c., and as the sand and fine gravel deposited in that part of the sluice has to be removed for the purpose of separating the gold from it, but little additional trouble would be involved in saving that portion of the sand and gravel for future manipulation in search of gems. It would not even be necessary that the labours of the miner should be interfered with, as the work of searching for gem stones in this refuse could be reserved for leisure hours. Of course to those miners who, being acquainted with the characteristics of gem stones in their rough state, could distinguish gems from other stones, the operation would be both simple and interesting, while those, who do not possess the requisite knowledge and experience to detect gems at sight, might collect all the transparent coloured stones and such white ones as present peculiar characteristics, and from time to time submit them to persons competent to express an opinion upon them. If the alluvium is reduced by the puddling machine, and the sand and gravel washed off by means of the cradle and tin-dish, the labour of searching for gems is so much lighter. In any case the search involves no expense, and so little labour as to fully justify the experiment on very many of our Gold Fields.

That the same may be said of our Tin Fields is evident from the following extract from a report furnished by Mr. Wilkinson to the Surveyor General in 1872:—"Sapphires of various colours are of common occurrence with the stream tin in Cope's Creek. Two diamonds are said to have been found in the Britannia Mine, Darby's Branch Creek, and sixty in one mine and forty in another mine on the Boro or Mead's Creek." Numerous other instances are cited by Mr. Wilkinson in his report herewith on the tin-bearing country of New England.

The reason assigned for discontinuing the search for diamonds in the Bingera district is that though they are sufficiently numerous, only a very small proportion of them are large enough for cutting, and there is great difficulty in finding a satisfactory market for the small ones. It is quite possible that since diamonds too small for the purposes of the jeweller are being applied to useful purposes, such as points of tools, &c., such a demand will be created and such a price obtained for them, as will render the working of the diamondiferous deposits at Bingera and elsewhere in this Colony highly remunerative. It is unfortunate that fair samples of these diamonds could not be secured for the collection of minerals now being made by this department, so that the attention of those who might become purchasers could be attracted. A jeweller in this city, who was largely interested in the question, says some 2,000 carats have been sent Home. That the weight of the diamonds ranged from $\frac{1}{32}$nd of a carat to 10 carats, and the value ranged from 3s. to £7 or £8 per carat.

The following paper on the Bingera Diamond Field, by Professor Liversidge, F.C.S., F.G.S., University of Sydney, late University Demonstrator of Chemistry, Cambridge, was read before the Royal Society, 1st October, 1873:—

In the following note I purpose giving a few facts concerning the recently opened diamond workings in the neighbourhood of the town of Bingera. Bingera is situated some 400 miles north of

Sydney, on the Horton, or, as it is more popularly termed, the Big River. This river runs into the Gwydir River, the Gwydir in turn losing itself in the Barwon or Darling River.

Being on my way last winter (June, 1873) to visit the tin districts of New England, I turned aside and availed myself of the opportunity to pay a hurried visit to the above diamond workings. The trip was not a satisfactory one, for, owing to the persistent rains and floods, travelling was at times quite impracticable, and at all times done under difficulties; hence, in the limited time at my disposal, all hopes of anything like a thorough geological examination of the spot had, to my great regret, to be relinquished. However, I was enabled to acquire a certain amount of information, which I venture to lay before you this evening, in the hope that it may not prove to be altogether devoid of interest and value.

But, in the first place, I may perhaps be permitted, *en passant*, to preface my remarks upon Bingera by briefly mentioning a few of the facts relating to the other and longer known diamond-bearing localities of Australia, but only so far as they throw light upon the Bingera deposits. For fuller information I must refer you to the Rev. W. B. Clarke's Addresses to the Royal Society of New South Wales in the years 1870 and 1872, and to the very complete account of the Mudgee diamond district, by Mr. Norman Taylor and the late Dr. Thomson, read before this Society in 1870.

Diamonds in Australia.

As early as 1860 the Rev. W. B. Clarke mentions the discovery of diamonds in the Macquarie River, but no information is furnished as to the conditions under which they were found, and it is not stated whether they occurred in the present river bed or in an ancient river drift.

But we have a full account of the geology of the diamond-bearing district detailed in the above-mentioned paper by Messrs. Norman Taylor and Thomson, and from it we shall see that the Mudgee and Bingera districts have many points of resemblance.

The Mudgee diamond workings are distant some 170 miles south of Bingera, on the Cudgegong River, which runs into the Macquarie River, and that again into the Darling River.

Diamonds were first discovered here in 1867 by the gold diggers, who neglected them for some time, but in 1869 they were worked pretty extensively. The localities lie along the river in the form of outliers of an old river drift, at varying distances from the river, and at heights of 40 feet or so above it. These outliers are capped by deposits of basalt, hard and compact, and in some cases columnar. This basalt is regarded by Mr. Taylor as of Post-Pleiocene age, but this has not been determined directly by any fossil evidence.

The great denudation which the district has sustained is at once apparent from the drift, together with its protective covering of basalt, having been cut up into these isolated patches or outliers.

The remains of the drift can still be traced for some 17 miles up the river, and in parts it still retains a thickness of 70 feet.

The patches which were worked, as enumerated in the above-mentioned paper, are as follows:—Jordan's Hill, 40 acres; Two-mile Flat, 70 acres; Rocky Ridge, 40 acres; Horseshoe Bend, 20 acres; Hassall's Hill, 340 acres. Total, 510 acres.

A peculiar deposit of crystalline cinnabar was found in one patch.

In the above localities the drift has invariably been met with in tunnelling under or sinking through the basalt, and in places where the basalt had been denuded away the drift has either disappeared or has been scattered over the neighbourhood.

No diamonds have been found in the river bed, except in places where the diggers have discharged the drift into the river when washing for gold.

The basalt when not resting on the drift frequently lies upon metamorphic shales, slates, sandstones, or greenstone.

The general formation of the neighbourhood is regarded as Upper Silurian, with overlying outliers of undoubtedly carboniferous age.

The rocks in the vicinity are nearly vertical, with a general strike of N.N.W., and consist of red and yellow coarse and fine grained indurated sandstones; thin white platy argillaceous shales; pink and brown fine-grained sandstone, banded with purple stripes; slates and hard metamorphic schists; hard brecciated conglomerate, containing limestone nodules, flint, and red felspar in a greenish silicious base. And with these occur dykes and ejections of intrusive greenstone.

The rocks are generally devoid of mica. For the most part the Older Pleiocene diamond-bearing drift is coarse and loose, but parts are cemented together into a compact conglomerate by a white cement of a silicious nature, sometimes rendered green by admixture with silicate of iron; in other cases oxides of iron and manganese have been the agglutinating agents. Diamonds were proved to exist in this solid portion by a special experiment of Mr. Taylor's.

The drift is chiefly made up of boulders and pebbles of quartz, jasper, agate, quartzite, flinty slate, shale, sandstone, with abundance of coarse sand, and more or less clay.

The quartz pebbles are white, like vein quartz, but often encrusted with films of iron or manganese oxides.

Many of the boulders and pebbles are remarkable for a most peculiar brilliant silicious polish, which is evidently not due to friction, since the cavities are equally well polished. Silicified wood is common, and coal has been found in the river higher up; also carboniferous fossils, such as Favorites Gothlandica and others.

Amongst the minerals associated with the diamond are the following (This list, we shall see, is almost identical with that furnished by Bingera) :—

1. Black vesicular pleonast. This mineral has not yet been found at Bingera.
2. Topaz.
3. Quartz.
4. Corundum.
   a. Sapphire.
   b. Adamantine spar.
   c. Barklyite.
   d. A bluish white variety, characteristic of Mudgee.
   e. Ruby.
   f. Rolled corundum, dirty white and pink.
5. Zircon.
6. Tourmaline.
7. Black titaniferous sand.
8. Black magnetic ironsand.
9. Brookite.
10. Woodtin; rare.
11. Garnets.
12. Iron, from tools.
13. Gold.
14. The Diamond.

The largest found was $5\frac{5}{8}$ carats = roughly 16·2 grains. The average sp. gr. was 3·44; and the average weight 0·23 carat, or nearly one carat grain each. The carat contains 4 carat grains which are equal to 3·16 grains troy.

The Newer Pleiocene drift afforded a few diamonds, and being derived partly from the older drift, its materials are somewhat similar; but in addition to the gems as enumerated, a few grains of osmiridium have been collected from it.

Diamonds have also been found in Victoria, but in no large quantities, and of but small size, but no report of their geological position appears to have been published.

We will now return to the more immediate subject of this note.

The Bingera Diamond Workings.

The diamond-bearing deposits at present undergoing development are some seven or eight miles, more or less, to the south of Bingera, and are situated in a kind of basin or closed valley amidst the hills; this basin is about four miles long by three wide, and is open to the north.

This, together with the surrounding district, is evidently of Devonian or Carboniferous age, but all attempts to procure fossils in order to verify this have hitherto failed. As before mentioned, the weather was too wet to allow me to make a proper search myself; in fact, it was only with very great difficulty that one could get about at all in the then state of the country. The weather was so thick from the pouring rain and constant mists that but meagre and unsatisfactory glimpses were obtained even of the country's general aspect. Nearly the whole of the basin seems to have been originally more or less covered with drift, parts of it having since been removed by denudation.

Running into the valley are various spurs of trap, which apparently cover portions of the drift; but at present this is only a conjecture, since the workings have not yet been carried on sufficiently far to show whether this be the case or not, neither by tunnelling under it nor by sinking shafts, but I hope soon to receive information upon this head, for when on the spot I suggested that a shaft should be sunk, which will decide the question. Should the drift be proved to pass under the trap, the known diamond-bearing area will be greatly increased. The probabilities are in favour that it does.

Both the trap and the drift have undergone much denudation.

The drift is said to be traceable along the course of the river for some (30) thirty miles.

The drift is the forsaken bed of some river, and in all probability that of the Horton.

The rock upon which the diamond drift rests, or the "bed rock" of the minerals, is an argillaceous shale. Outcrops of this are seen in one or two places, but no good section is shown.

In other parts of the ground we see a compact, rather small-grained silicious brecciated conglomerate, strongly agglutinated together by a ferruginous cement; occasionally the pebbles incorporated in this conglomerate are of rather large size.

In one part of the field the junction of the conglomerate with the argillaceous shale is clearly shown in the cutting formed by a small gully.

Both the shale and conglomerate beds appear to have undergone much disturbance; and at this particular spot diamonds are said to be plentiful on the conglomerate but not on the shale. The surface of the shale is here free from drift, but the conglomerate does not appear to be quite free from it. The miners regard the conglomerate as being of itself diamond-bearing, but this has not been put to any absolute proof.

Up to the present all the diamonds have been found within a foot or so of the surface, in fact just at the grass roots. In no case have the workings been carried to greater depths than two or three feet; in some parts examined the drift itself is not thicker than that.

In the former sinkings made by the gold diggers diamonds have occasionally been met with at depths of 60 feet, or even more; but, as the men were working for gold, no great attention was paid to the diamonds, and it is quite likely that they fell in from the surface.

The method employed in the search for the precious stones is very simple:—The drift is stripped off and carted to the puddling-machines, where it undergoes a great diminution in bulk by the removal of the clay and fine sand; the large pebbles are then screened off, and the clean gravel remaining is passed through one of Hunt's diamond-saving machines. But since this apparatus depends upon the principle of separation by difference in specific gravity, it does not perhaps afford the best method which could be devised; it may answer well enough for gold and other bodies of very high specific gravity, but must certainly answer very imperfectly for diamonds, on account of their comparatively low specific gravity, viz., 3·4 to 3·5, which is nearly equalled by most of the accompanying minerals, and exceeded by some.

I should be inclined to recommend the methods employed in Africa and Brazil, since they would probably prove more efficacious.

We may now pass on to consider in more detail the mineralogical nature of the drift, or "wash-dirt" as it is termed by the miners.

From Messrs. Westcott and M'Caw's claims I obtained three different specimens. See samples Nos. 1, 2, and 3.

Wash-dirt No. 1.

This is a pale brown clay, binding together well-rolled pebbles, subangular and angular fragments of variously coloured jasper, red, green, brown, &c. Also black flinty slate, tourmaline, argillaceous sandstone, and shale, &c.

Wash-dirt No. 2.

This is rather darker in colour than No. 1, and the clay is more tenacious, the contained pebbles are of much the same character; the clay has a brecciated structure, and differs in colour in parts, fragments of it being nearly white. On the spot, when freshly dug out, portions of the clay are of a bright green colour, due to the presence of a ferrous silicate, which, by exposure to the air, absorbs oxygen, and passes into the reddish ferric silicate which imparts the red colour to the clay.

Wash-dirt No. 3.

This kind contains a larger proportion of pebbles than either No. 1 or No. 2; it is of a light colour, and much less indurated, being of a sandy nature. This also contains pebbles of argillaceous shale.

Unrolled blocks of the bed rock are met with in all the drift. In all three we find occasional minute crystals of selenite, probably of very recent origin.

During the process of extracting the diamond from the wash-dirt, the material is sized as it passes through the machines ; but as it is hardly necessary to consider these sands and gravels separately, it will be as well to consider their constituents merely, irrespective of the size, since they all contain nearly the same minerals, although not in the same proportions ; but as the large pebbles and boulders which are removed immediately after the stuff is puddled do differ from the finer parts considerably, we shall take them by themselves.

Pebbles and boulders.

These consist of masses of red, green, brown, and other coloured jaspers ; white quartz, common agate, black flinty slate ; fine sandstone, into which manganese and iron oxides have infiltered, leaving dendritic markings between the joints. Many of the pebbles are also coated externally in the same way. Nodules of magnesite and concretions of limonite or brown iron ore, of concentric structure,—some of the magnesite still showing the limonite *in situ*. Rolled masses of hard compact brecciated conglomerate, often containing much manganese in the cement ; masses of silicified wood (but this is not very common), cacholong, and greenstone. The rolled masses of sandstone, and especially the argillaceous sandstone, often assume long finger-shaped forms, and are accordingly termed "finger stones" by the diggers. The pebbles are not polished, as at Mudgee.

The list of gems, stones, and other minerals accompanying the diamond, includes the following :—

1. *Tourmaline*, or "jet stone," of the miners, occurs as rolled prisms, usually from a $\frac{1}{4}$ to $\frac{3}{4}$ inch long. They usually retain the trigonal section, but sometimes no trace of crystalline form is left, and they appear merely as more or less rounded black pebbles, often with a pitted surface, totally unlike the usual appearance of tourmaline ; the blow-pipe decides their character at once, for they intumesce before it and in other respects answer to the well-known tests. These "black jet stones" are invariably found with the diamond, and are regarded by the miners as one of the best indications of its presence.

2. *Zircon* occurs in small crystals of red and brown colours, also nearly colourless, but more commonly as rolled pieces of a brown shade. A cleavage plane is usually to be seen.

3. *Sapphire*, generally in small angular pieces and usually of a pale colour ; in many the blue tint does not overspread the whole of the fragment. The *Ruby* is present, but very rare. One fragment showed the faces of an acute hexagonal pyramid and basal pina-

coid. The lower half of the crystal had been fractured; it was of a red colour, but possessed a purple-coloured central mass. The fragments of sapphire are far less in size than those found at Mudgee and in New England, and far less rolled; the major part often appears to have undergone no rounding at all—thus presenting a broad distinction between the gem sand from the two places. A little corundum is found.

4. *Topaz*, as rounded fragments, and sometimes with rough crystalline outline. They are generally of a dull yellowish colour, colourless and transparent, small in size, and often apparently freely fractured.

5. *Garnet*, in small, rough-looking ill-formed crystals, of a dull red colour.

6. *Spinelle:*—Not very common, generally in small red or pinkish fragments.

7. *Quartz:*—Small prisms, capped with the pyramid, more or less rolled, transparent, and of a pale dirty red, also smoky; also small jasper pebbles, &c., &c. Amongst the jasper pebbles are some of pale mottled tints of yellow, pink, drab, brown, bluish grey, &c.; these are termed "morlops" by the miners, and are regarded by them with much favor, as they say they never find one of them in the dish without diamonds accompanying it. Their average specific gravity, taken from a large number, is 3·25. As this is nearly the same as that of the diamond, we can readily understand their being found together. Many must be lost in the washing processes. They are oval in form, smooth, and rarely exceed a quarter-inch in length. The miners can give no origin for the name, and it does not appear to be mentioned in any works on mineralogy, &c.

8. *Brookite:*—Small flat fragments—very rare.

9. *Titaniferous iron:*—Rather common.

10. *Magnetic iron ore*, in small grains, showing an octahedral form under the microscope, coated with hidrated sesquioxide of iron, easily removed by the magnet. Gold in small particles was often found attached to the grains of magnetite.

11. *Wood tin:*—Rare; in small rolled particles.

12. *Gold:*—Fine grains and scales, present but in small quantity, and the greater portion attached to the magnetite; hence the magnet was found the most ready means of removing it.

13. *Osmiridium:*—In small brittle plates; rare.

14. *The Diamond:*—As already stated, they are for the most part small in size. Some are clear, colourless, and transparent, while others have a pale straw-yellow tint. One or two dark ones, very small, have been seen; also a greenish one. The sp. gr., as deduced from nineteen specimens, is 3·42 (the Mudgee being 3·44).

In some the crystalline form is well and distinctly shown, but others possess very much rounded faces. Some of the best crystals were those of the triakis-octahedron, the triakis-tetrahedron, the octahedron, the tetrakis-hexahedron (or four faced cube), and the hexakis-octahedron.

No fractured specimens have been detected, but it is rather common to find them with very much pitted surfaces, and with internal black specks.

One of the Companies, when prospecting the ground, found the drift to yield as follows :—

6 loads yielded	41	diamonds
4½ „	143	„
6 „	88	„
6 „	125	„
6 „	163	„
6 „	89	„
Refuse from machines, &c....	41	„
31½	690	diamonds,

or an average of 20 diamonds per ton of stuff, regarding the load as equal to one ton. The above were obtained by the Gwydir Diamond Mining Company.

The following is an account of the number obtained by Messrs. Westcott & M'Caw from the Eaglehawk claim, up to August 26th, 1873 :—

400	diamonds, weighing	192	grains
420	„ „	199	„
310	„ „	153	„
200	„ „	109	„
350	„ „	150	„
1,680	„ „	803	„ troy.

And, as examples of the number obtained per load of stuff, the following may be cited :—

5 loads yielded 86 diamonds, weighing 32 grains.
8 loads yielded 68 diamonds, weighing 30 grains.

Up to the present no large diamonds have been found, the largest hitherto met with being one only of 8 grains

1 of 4 grains
6 of 3 „
85 of 2 „
1,587 of less than 2 grains.

No mention is made of the kind of drift from which the above quantities were obtained; they, however, afford an opportunity of roughly estimating the yield.

"It is reported from Bingera, in the *Tamworth News* of September 26, that Mr. Gardiner has obtained 115 diamonds, and that the Gwydir Company are progressing vigorously. The Giant's Knob is rich in gems, the yield averaging about 140 to the machine full, when the dirt is taken from the diamond drift.

"A correspondent of the *Tamworth Examiner*, on the 12th instant, states that there have lately been large finds of diamonds in the district of Bingera. The Gwydir Diamond Company have prospected now twenty-one pieces of land, nineteen of which have proved to be more or less diamond-producing soil, containing Grupiara or alluvial deposit, whose surface shows it to be the unused bed of a stream or river: Burgalhas, small angular fragments of rocks, bestrewing the surface of the ground; Cascalho, fragments of rocks and sand mixed up with clay and forming the bed of a river; and Takoa Carza, which are the above materials cemented together into a conglomerate mass. All the above, however, are known by the generic name of Cascalho. The masses of stones themselves, which rarely exceed a cubic foot in size, contain itacolumite jasper, and perdots and granite. These are the known indications of the whereabouts of diamonds as trusted to and found to be correct both in the East Indies and the Brazils. The nineteen successful prospects of the Gwydir Company have produced each on an average thirty-five diamonds to every six loads (of one ton) of wash dirt, and they have now by them some 11,000* glistering pebbles, ready to transmit to Amsterdam, Paris, or some other European continental market; and are at present making extensive arrangements for the formation of three more dams and puddling apparatus on other parts of their land where good supplies of water are to be found. He also gives the following as the find of Messrs. M'Caw and Westcott: Up to the week ending July 12, 100 diamonds; up to the week ending July 19, 113 diamonds; up to the week ending July 26, 119 diamonds —total 322"—*Herald*, August 21, 1873.

The only minerals found at Mudgee which have not yet been discovered at Bingera are cinnabar and vesicular pleonaste. Bingera in turn seems to possess one or two characteristics, such as the magnesite containing the nodules of limonite—these are perfectly spherical at times—and the "morlops" form of jasper. As these are nothing more than small jasper pebbles, careful search would probably prove their presence in most river drifts containing rolled jasper.

From the foregoing we see that the diamond in Australia is associated with sandstones, shales, conglomerates, and trap-rocks; and, perhaps it would not be amiss just to see if this be the case in other countries. Thus, in the Brazils, in Bahia, the matrix

* (1,100 query?)

of the diamond is said undoubtedly to be a tertiary sandstone. Burton,* in his book on the Highlands of Brazil, states that it occurs in itacolumite, a metamorphosed palæozoic rock; but this statement requires confirmation. This, however, is known indisputably,—that they occur in the alluvial drifts of various kinds similar to those of New South Wales.

Diamonds found on the Cuddapah Hills in India are stated to occur in a conglomerate, and between Saugor and Mirzapore in a solid sandstone, and also in a ferruginous conglomerate; and in a gravel at Cuddapore, containing pebbles of trap, granite, schist, quartz, jasper, sandstone, and also of the neighbouring limestone; basalt also is found near by.

And at Banganapilly the diamond is said to have been found in a sandstone, together with corundum and magnetite, as well as in breccia, and the slates there are flinty. The district of Kumarea and Bridgepore is conglomeratic, and associated with sandstone beds. Other diamond-bearing localities of India also are conglomeratic.

In Russia too conglomerates seem to be the present receptacle of the diamond; iridium is there associated with it.

Borneo—again here we find it in a conglomerate containing quartz, &c., and associated with gold, platinum, and osmiridium. Then too, in Africa, they are found in a drift, and usually within a few feet of the surface, from 3 to 9 feet, and rarely down so far as 30 feet.

Here, again, one of the main features of the district is the presence of sandstone, either of Upper Silurian or Devonian age; trap is also present, and a conglomerate or breccia, containing boulders of granite, gneiss, mica schist, porphyry, sandstone, jasper, slate, agate, &c.

In conclusion, we are still as much in the dark respecting the origin of the diamond or even its true matrix, for no good proof has yet been offered on this question, as we have seen in nearly all cases it occurs in an ancient river drift, and is usually associated with sandstones, conglomerates, and trap rocks; neither do we know the matrix of the sapphire, zircon, &c., which are usually much rolled, as if they had been borne a great distance. The sapphires from Bingera seem to have undergone but little alteration, and consequently have not travelled far, so that perhaps we may soon light upon their source and that of the diamonds simultaneously. Bingera certainly seems the most hopeful locality to elucidate this point of any at present known.

Before closing this paper, I must express my obligations to Messrs. Westcott and M'Caw, and to Mr. Dougherty, of the Gwydir Diamond Co., for their great assistance in procuring and

---

* See vol. II, p. 144.

sending me suites of specimens illustrating the various rocks and minerals of the diamond workings, and wish them success in their endeavours to open up this industry, which I hope will prove to be a new source of wealth to New South Wales. And this result appears to be highly probable, since the whole of the above-mentioned valuable finds have been made by the exertions of but a few, perhaps not more than five or six, workers.

APPENDIX.

REPORT ON THE DISCOVERY OF DIAMONDS AT BALD HILL, NEAR HILL END.

University of Sydney,
December 5, 1873.

To the Hon. the Minister for Lands,—

Sir,

In reply to your request of the 2nd instant, I have the honor to furnish you with the following particulars relating to the mineral specimens from Bald Hill, near Hill End, which accompanied your communication.

Diamonds—Three in number; the largest of them is in the form of a six-faced octahedron, rather flattened, owing to four of the groups of faces being more highly developed than the remaining four. The faces and edges are rounded somewhat, but this has not been caused by attrition; diamonds often appear as if water-worn, but in reality this is seldom the case; the rolled and water-worn appearance is due to the fact that the diamond usually crystallizes with curved faces and rounded edges. It is clear and colourless, and perfectly free from all visible internal flaws; the surface is likewise free from flaws; but scattered over some of the faces are a few minute and insignificant triangular markings, but these are quite superficial and will disappear during the ordinary process of cutting. It possesses a specific gravity of 3·58, and weighs 9·6 grains (Troy), *i.e.*, a little over three carats. It is generally calculated that diamonds lose one-half their weight during the process of cutting and polishing; and their true value cannot be ascertained until this has been done. The diamond next in size possesses the same crystallographic form as the one above mentioned, but is not so much compressed. It has a weight of 4·5 grains (Troy), or nearly one and a half carats. It has a chip on one edge, and contains a speck of foreign matter. It is a straw-colour. The smallest diamond weighs about half a grain; it has the form of a six-faced tetrahedron, and possesses a high lustre, but is rather off colour.

Accompanying the diamonds were two small specimens of gem sand.

GEM SAND NO. 1.

In this the following substances were found to be present:—

I. *Corundum*—When blue this is known as the *sapphire*, and when red as the *ruby*.

(a) Common Corundum—Present in small fragments of bluish, greenish, and grey tints.

(b) Sapphire—In small particles of a blue colour, some so dark as to appear almost black, and others very light. Some of the fragments still show their crystalline form, viz., a hexagonal pyramid, but most of them do not, and are either much rolled, subangular, or angular in their outline.

The ruby is absent, but probably would have been present had the sample of gem sand been larger.

II. *Zircon*—Plentiful, usually in the form of much rolled pieces. Generally of a brown colour, sometimes red and at others nearly colourless. The small and nearly colourless crystals possess a very high lustre, almost equal to that of the diamond, so that they might readily, without careful examination, be mistaken for that gem.

III. *Quartz*—Usually as small, well rolled grains, either colourless, milky, or yellowish. Sometimes as hexagonal prisms, capped with the hexagonal pyramid. Jasper of various colours, such as red, yellow, grey, also occurs, together with black grains of flinty slate.

IV. *Rutile*—In angular fragments, still showing traces of crystallization. Distinguished by its brown colour and metallic lustre, and by the presence of numerous fine striæ on the faces of the prism. It very much resembles tin stone in appearance. In composition it consists of titanic acid.

V. *Brookite* also occurs. This is another form of titanic acid. Rutile crystallizes in striated tetragonal prisms, whilst this crystallizes in tabular forms belonging to the rhombic system. It is present in small quantity, in the form of flat irregular plates, brown or grey in colour.

VI. *Topaz*—Present in small rolled and angular fragments, colourless, and in pale tints of yellow and greenish blue. The latter coloured topaz is often erroneously termed the aquamarine.

VII. *Beryl or Emerald*, doubtful, but one or two very small fragments resembling it.

VIII. *Garnet*—Small, rough, common garnets, of no value.

IX. *Tourmaline*—A few rounded pieces, but none showing the crystalline form, which is that of a three-sided prism.

X. *Gold*—Present in the form of scales.

GEM SAND NO. 2.

This consists of larger grains than No. 1; in fact they are small pebbles.

I. *Quartz*—Present principally in the form of jasper, of various colours, red, brown, green, yellowish, &c., &c.; also variegated. Colourless and yellow quartz pebbles are also found, together with black pebbles of flinty slate.

II. *Corundum*—Present as common corundum, and as the sapphire.

III. *Brookite*—Same as gem sand No. 1, only in larger pieces.

IV. *Topaz*—Clear and colourless; also tinted.

From the foregoing it will be seen that the Bald Hill gem sands very closely resemble those from Bingera and Mudgee.

None of the gems contained in the parcels submitted to me, with the exception of the diamonds, are of any commercial value, except for grinding and polishing purposes. Still, they are of great value as indications, for where such occur there is every prospect of finding others of larger size and better quality.

An examination of the original "wash dirt" or "drift" might yield valuable information, and larger samples of the gem sand will probably be found to contain such minerals as iridium, titaniferous iron, tin stone, magnetite, &c., like the Bingera gem sand.

I may perhaps mention that in 1867 a brilliant of the first water and without flaw, weighing one carat, was worth about £20; if weighing one and a half carat, about £45; and if two carats, about £80, and so on; but since that time the prices have probably undergone much change. According to the September number of the British Trade Circular, the prices ruling for Cape diamonds, uncut, are in proportion much lower than the above.

I return the diamonds and gem sand per bearer.

I have the honor to be,

Sir,

Your obedient servant,

ARCHD. LIVERSIDGE.

MINERAL EXHIBITS.

That the collection made by this department is not larger and more complete is due to the fact that, with the exception of the valuable collection of fossils, &c., &c., made by the Government Geologist, and the fossils and samples of Coal, &c., collected by the Examiner of Coal Fields, the whole of the specimens (embracing samples of ores, &c., from nearly every district in the Colony) have been collected since the commencement of the present year. The readiness with which owners and managers of mines have contributed samples, and the energy displayed by the officers of this department in collecting and transmitting them, are deserving of the highest praise.

It is the intention of the department to obtain from each district a complete collection of samples of every mineral possessing an economic value found therein, and to arrange the collection in such a manner as will enable the public to see at a glance the mineral products of every district in the Colony. The making of such a collection will of course occupy some time, but no efforts will be spared to make and arrange it as speedily as possible.

The following are notes by C. S. Wilkinson, Esq., Government Geologist, on the Geological and Mineralogical Collection exhibited at the Metropolitan Intercolonial Exhibition, 1875:—

The geological collection which I have had the honor to arrange for the Department of Mines at the Intercolonial Exhibition comprises upwards of 1,000 specimens illustrative of the geology and mineral wealth of New South Wales.

Owing to the short time at my disposal for this work, and to the recent establishment of the Geological Survey, this collection is not so complete as it might have been. Nevertheless, with the assistance afforded by the Wardens, Mining Registrars, and other officers of the Department of Mines, the Surveyor General and others hereafter mentioned, the fossils and minerals I have now arranged, fairly represent the principal geological formations with their mineral resources, and cannot fail to show their scientific and economic importance.

Though an outline geological sketch is here necessary for reference, the present paper, I would premise, must be regarded rather as a general description of the collection exhibited than as a full exposition of the Geology of New South Wales. The latter will be found in the 3rd edition (published herewith), of *Remarks on the Sedimentary Formations of New South Wales*, by the Rev. W. B. Clarke, M.A., F.G.S., whose extensive geological explorations in Australia command his authority on the subject.

The order of the formations may be stated as follows:—

Post Tertiary	*Recent.* *Pleistocene.*
Tertiary or Cainozoic	*Pliocene.* *Miocene.*
Secondary or Mesozoic	*Triassic?*
Primary or Palæozoic	*Permian.* *Carboniferous.* *Devonian.* *Upper Silurian.*

The above classification must, to a certain extent, be considered only provisional; it cannot be otherwise until the mutual relation of the various members of each formation has been ascertained by actual survey.

RECENT.

Under the term *Recent* are included all those deposits of gravel, sand, loam and mud, which are accumulating at the present time. Such are the loam and clay of surface soils, resulting from the decomposition and disintegration of the underlying or neighbouring rocks; the sand and gravel swept along by running streams; the earthy mud and *debris* spread over swampy and low-lying lands by floods; and the ever-shifting shingle-beaches and sand-dunes piled along the sea-coast by the action of both waves and wind. As would be expected, amongst these various accumulations are found mingled remains of animals and plants of species now living, with human bones, *mogos* or stone tomahawks, bone needles, flint chips, and other implements of the aboriginal tribes now fast dying out.

The economic importance of the recent deposits is very great. In them gold was first discovered, and, up to the present time, they have been a source of employment to thousands of gold-miners. In New England they have also afforded the chief supply of the alluvial or stream tin ore, which, during the past three years, has become such an increasingly valuable addition to our mineral exports. A description of the various modes of occurrence of the stream tin will be found in my Report (1873) on the Geology of the Tin-bearing Country near Inverell. Through the courtesy of the Surveyor General, Mr. P. F. Adams, under whose directions my geological examination of that part of New England was conducted, the various descriptions both of stream and lode-tin ores, which I then obtained, are now exhibited. But, whilst these represent the mineral products of the Inverell district, we have also a number of good samples, especially of stream tin, collected by Mr. George Gower, Mining Registrar, from the mines of Vegetable Creek; and these have been supplemented by others obtained by Mr. J. H. Butchart, from various parts of New

England, including over a ton weight of solid lumps of tin ore, partly waterworn, from Beardy Falls. The Glen stream tin is obtained in some quantity in the southern parts of the Colony, near Albury, on the Murray River, and at Tumut. A fine sample of *Toad's-eye* tin from Grenfell is also shown. So that the present collection of tin ores now exhibited by the Department of Mines will afford evidence, not only of the excellent quality of the ores, but also of their extensive distribution in the highlands of New England.

Mr. J. Buchanan, Warden of the Northern Mining District, states, in his annual report for 1874, that "the tin fields, especially at Vegetable Creek and Cope's Creek, continue to yield steady returns, and the industry after much fluctuation, may be said to be thoroughly established. Townships at both of these places are in the most flourishing condition, and there seems every prospect of many years elapsing before the ground actually in process of working will be exhausted. I estimate the population on the tin fields at 2,000 men. The existence of valuable reefs was satisfactorily proved some years ago, but as yet nothing has been done towards their development. I contemplate, however, increased activity in the working of the tin mines during the ensuing year, and I may state that, from personal inspection and experience, I am convinced that a wide field is open to the labouring classes upon them."

Pleistocene.

The Pleistocene deposits chiefly embrace those alluvial flats which are found more or less extensively along the course of almost every river and creek. They occur often at considerable elevations above the present running stream, forming terrace-like plateaux on the sides of the valleys. These terrace-drifts indicate the depth to which the valleys had been eroded at the time they were deposited; they are, as it were, the remaining flood-marks of the streams which, during a lengthened period, gradually excavated the existing valleys—streams which swept away the disintegrated earth, and redeposited it over distant lowlands. The origin of those extensive loamy plains of the Riverina and the Western Districts may be attributed to the enormous atmospheric denudation, which chiefly produced the rugged features of the mountain ranges.

During this period lived several species of gigantic marsupials; amongst others have been described the extinct genera Diprotodon and Nototherium; the Sarcophilus ursinus, or Tasmanian Devil, some gigantic Kangaroos, together with the Dinornis or Moa. The bones of many living species of animals are also entombed in the accumulations of this age, to which belong the deposits in the Limestone Caves of Wellington, &c.

The Pleistocene drifts exceed in importance those of the recent, being of far greater extent, and equally rich in alluvial gold and tin. In the diggings districts they are seldom more than thirty feet thick. Mr. Chas. de Boos, Warden of the Gold Fields, Southern District, in his report on the Araluen Valley, where the stripping is from twenty to thirty feet, says:—"To give an idea of the work carried on, I may mention that one claim—that of Herbert and party—employs at the present time seventy men and forty-five horses, whilst they have three powerful engines on the ground. Only two other extensive claims are at work, and these employ somewhere about fifty men each and thirty-five horses, besides the engines and machinery for pumping."

In New England the drifts contain rich stores of stream tin, and are now extensively mined. They will continue to afford profitable employment for many years, after the more easily worked recent drifts have been exhausted.

Pliocene.

Most of the present surface features have had their origin during the Pliocene period. The Recent accumulations are those now in process of formation; and we have seen that the Pleistocene drifts, &c., include those which were deposited whilst the present valleys were being eroded; and, in pursuing our investigations further, we observe that the formation through which many of these valleys have been excavated is basaltic trap, and that this volcanic rock in its molten state has filled in and covered up *older* river courses, which existed and had their origin in the Pliocene period. We again further notice that these older valleys have cut through thick strata of gravels and clays, which had been deposited since the Miocene period. Between the Pleistocene and Miocene epochs, we have thus, at least, two distinct series of deposits, which may be classed as Upper and Lower Pliocene; the former embracing those old fluviatile drifts buried in places beneath basaltic lava; the latter, those wide-spread auriferous gravels and clays which appear to be of marine origin. Our geological history, therefore, reveals this important fact,—that after the Miocene period, this continent was submerged beneath the sea, when the Lower Pliocene marine deposits were formed; and again raised and subjected to rain and other atmospheric denuding agencies, whose well-known influence gave rise to that general system of valleys which now furrow the slopes of the Cordillera. There is no evidence to show that the land has been again submerged since the Lower Pliocene times; on the contrary, the present physical features rather indicate that the atmospheric denudation which then commenced has continued to the present day, disturbed only by several volcanic emissions, which chiefly at the close of the Pliocene epoch occurred. About this time, in many parts of the continent, numerous volcanoes burst out, their intermittent streams of lava pouring into the adjacent valleys, overflowing river courses and the bordering forests, and in their onward progress spreading far and wide, to form those rich basaltic downs of the western districts. These features are frequent both in New South Wales and in Victoria, and it is of great importance that they should be carefully investigated, not only with the object of elucidating geological history, but also a correct understanding of the facts they reveal will greatly facilitate the progress of mining the deeper alluvial leads. Those buried old river beds now constitute the so called deep leads. As yet they have been tested only near their sources, where they formed but the tributaries of some larger river-channels not yet discovered, but the existence of which is obviously certain. Mr. T. A. Brown, Warden at Gulgong Gold Fields, has expressed his fixed belief in the future discovery in his district of rich deposits underlying the basaltic formation at greater depth than has hitherto been reached. And Mr. Fred. Dalton, Warden, in his annual report for 1874, gives the following interesting particulars respecting the deep leads in the Lachlan Mining District: "On several leads discovered and worked during the past year it has been proved that the deepest ground is not auriferous; shafts have been sunk to depths varying from 140 to 160 feet, and bottomed upon a wash-dirt that differed in no respect from that obtained in adjoining claims at depths of from 80 to 90 feet, except that the former did not contain any gold, while the latter yielded from 10 to 15 dwts. per load. These watercourses run in a parallel direction, and are not 100 yards distant from each other."

Such not uncommon occurrences may have originated in various ways. For instance, at the particular time of the deposition of the gold-bearing drift, the old stream may have been wearing away and distributing as drift some auriferous bed of rock or quartz reef which had no great depth; consequently, when the river had eroded its channel to a lower level, the underlying rock it then acted upon may have been non-auriferous, and therefore the drift derived therefrom would be so also. Again, in silurian country there have been frequently noticed belts of non-productive rock, running parallel and alternating with rich tracts; these then under certain circumstances would also afford an explanation of the above-mentioned facts. I could also cite instances where alluvial drifts had derived their richness from the bed of an old auriferous lead which had been cut through, and its contents redistributed: any further deepening therefore of the valley in the rock below the old lead, would cease to produce payable drift; in this manner also may it often occur that the drifts left at various heights in the same valley will differ in richness.

Mr. Dalton continues—"A band of argillaceous limestone of an unascertained depth and width, in which no fossils have been as yet discovered, extends from the bank of the Lachlan River due north, and intersects the Lachlan and Billabong Gold Fields, near the centre, cropping up on the limestone ridges about one mile to the west of Forbes. It again appears in the midst of surfacing at the head of one of the five tributaries of the "Bushman's" Lead, reappears near the "Dayspring" Reef, where its eastern margin is intersected by trappean dykes, and again comes to the surface to the northward of the "No Mistake" Lead, whence it passes onward through untried country.

"The most productive auriferous leads, both in the Lachlan and Billabong Gold Fields, have been opened on the flanks of the belt of limestone referred to and the leads now being worked, to the south of the Goobong Creek, follow the same course, and are similarly situated with respect to it. Nuggets of gold weighing from 2 to 9 ounces are frequently obtained from these leads. In the centre of the surfacing previously mentioned, a cavernous opening was discovered in the limestone, about 60 feet in width and 120 feet in length, the walls on all sides being perpendicular; this cavity was filled with the ordinary wash-dirt which had been drifted into it, and has been worked from the surface to a depth of 50 feet with payable results; as yet no bottom or horizontal opening has been reached. This cavern must at some former period have received the drainage of a wide valley.

"The alluvial leads discovered and worked up to the close of 1873 are nearly exhausted; their total yield was about 50,000 ounces of gold."

Five leads lately opened on the south side of Goobong Creek seem to be tributary to a main channel, as yet undiscovered. Some of these leads yield from 3 to 13 dwts. per load of wash-dirt; but on M'Guigan's Lead on the north side of that creek, at a depth of 103 to 115 feet, 1 to 2 ounces of gold per load is no uncommon return from the puddling machine. This lead is the most important discovery during 1874. The "bottom" is a pipe-clay, with occasional patches of compact claystone, sandstone, and decomposed diorite. The fine nugget of gold (hereafter mentioned) weighing nearly 23 ounces was obtained from this lead.

"From the Great Northern Block four men obtained 800 ounces of gold; six men from a claim, in which there is still eighteen months work, 600 ounces; four men, for four months labour, 450 ounces; and four men, in return for four months work, 350 ounces.

"During the past year 126,000 loads of auriferous drift have been passed through the puddling machines, at an average cost of 3s. 3d. per load, or a total cost of £20,507 10s. The average yield was 7 dwt. 14 grs. per load, and the total produce of the alluvial mines 47,868 ounces 2 dwts. 22 grs., the increase being considerably in excess of 100 per centum on the production of 1873."

It has been the experience in New South Wales and in Victoria, that the productiveness of these auriferous leads depends much on the underlying geological formation. Alluding to this fact, in reference to the tertiary drifts in Victoria, Mr. A. R. C. Selwyn, late Director of the Victorian Survey, and now Director of the Geological Survey of Canada, states—"That though they are richly auriferous when chiefly derived from and resting directly on the Silurian rocks, they gradually cease to be so when they become underlaid by rocks newer than Silurian; and thus the points at which the respective leads will die out and become unprofitable will probably be determined by the position in the several districts of such lines of junction." (*Notes on the Physical Geography, Geology, and Mineralogy of Victoria*). In New South Wales, Lower Palæozoic are the underlying formations for several hundred miles westward from the Cordillera; and as the conformation of the country is not unfavourable for the continuation in that direction of the known old leads, we may reasonably infer that their development is at the present day only in its infancy; and that as many of these ancient river-beds will be found to unite and increase in size the further westward they are followed, the scope they will afford for future mining enterprise will be practically unlimited.

From these auriferous leads vegetable fossils have been obtained, including trunks and branches of trees, with leaves and fruits. Specimens of the latter, obtained by Mr. Farr, Mining Registrar, Bathurst, I lately submitted to Baron Ferd. von Mueller, C.M.G., Government Botanist of Victoria, who has very kindly favoured me with the following description of them, which will be read with interest, inasmuch as the Baron shows that, amongst these fossils he has not only identified species characteristic of the upper pliocene auriferous leads in Victoria, but has also discovered a new genus—"*Rhytidocaryon*."

Rhytidocaryon.

"Fruit spherical or slightly ovate, not distinctly dehiscent, one-seeded, with an oblique basal or slightly lateral attachment, woody or bony, externally wrinkled and somewhat tuberculate. Septum large, placenta-like, erect or slightly ascending from the bottom of the cavity, consisting of two portions, which are smooth, turged, oblique, ovate, or sometimes broadly clavate or roundish, always more or less contracted at the base, mutually connate at the middle, rounded at the edges, broadly adnate to the lateral parts of the cavity, free from its summit. Seed cylindrical, bent around the placental or septal protrusion, oblique orbicular or ovate hippocrepical in outline, with a marginal furrow. Testa, thin, brittle, smooth.

Rhytidocaryon Wilkinsonii

"Beneree, under basalt at a depth of 110 feet; Edwd. Farr Esq.; communicated by C. S. Wilkinson, Esq. Found also between Carcoar and Orange, by the Rev. W. B. Clarke, M.A. Fruits, which constituted probably separate carpels of a tricoccous fructification from two-thirds to rather above 1 inch long, externally uneven from somewhat irregular slightly concentric ridges, which are often broken up into short tubercules, approaching in roughness somewhat to those of Phymatocaryon Mackayi, probably covered originally by a pulpy pericarp, which in decay would early perish, thus the nut-like covering constituting a putamen or endocarp; a very faint cleavage at the base, but no trace of valvular dehiscence; septal process from less than double to nearly triple the width of the walls of the endocarp, except the base and back free from the cavity. Seeds (in all specimens under examination) perished, but their form recognized from the space left for their reception between the dissepiment and the inner faces of the endocarp; remnants of the testa not showing any indications to intrusions into the albumen. The latter and the embryo unknown.

"This new fossil, so far as I can judge from the material transmitted to me, brings before us for the first time with certainty a member of the Menispermeæ among the vegetation of bygone creations, inasmuch as of this order hitherto only the altogether doubtful genus Mac Clintockia (*Heer die Fossile Flora der Polarländer*, 114-116; *Schimper, Traité de Paléontologie Végétale*, III, 83-84 pl. XCVIII) became palæontologically recorded. Unacquainted as we are with the flowers and the embryonic characters of the fruit, we must regard it unsafe to place this into any of the numerous genera of Menispermeæ, distinguished mainly by their floral organization and the inner structure of their fruit; but the endocarp and septal protuberance show some resemblance to the South Asiatic genera Hypserpa (Miers, in Annals of Natural History, sic. ser. VII, 40), Limacia and Nephroica (*Laureiro Flora Cochinchinensis*, 620 *et* 692), and the East Australian Sarcopetalum (F. M., Plants indigenous to the Colony Victoria I, 27, pl. III, suppl.) The putamen, however, is more rough than that of any of these genera, and indeed conspicuously thicker than that of any living menispermaceous plant known to me, while in its great size the fruit of Rhytidocaryon shows only similarity (and in this respect merely) to Hæmatocarpus (Miers's Contributions to Botany, III, 324, t. 134). The leaves are unknown. It is probable that the plant yielding these fruits formed, like most of the menispermaceous order, a climbing shrub.

Spondylostrobus Smythii.

"The fruits of this tree are rather variable in size and shape, but preserve throughout a long series of varieties the cardinal characteristics of this extinct genus of Coniferæ, of which as yet but one species became known, described by me in Mr. R. Brough Smyth's Reports of Mining Surveyors and Registrars for 1871. A short notice of this fossil appeared also in the London *Geologic Magazine* for March, 1871. We had this identical plant hitherto from Nintingbool, the Tangil, Beechworth, and also from Orange in New South Wales among pliocene drift. This tree must therefore have occupied a vast area at that period. As foliage of pine-like trees is preserved readily in a fossil state, you may succeed in discovering the leaves, and also get the flowers of this conifer, for which hitherto on other spots a search has been made in vain. An excellent lithographic illustration of the Spondylostrobus accompanies Mr. Smyth's Report, above quoted. This fossil, together with the following, is indicative of auriferous strata.

"No. 4 is the fruit of *Penteune Clarkei*, of which a description and lithogram was offered in the Mining Reports of the Victorian Department in 1873. Also, of this tree, which was a companion of the Strongylostrobus, but seemingly less frequent than the latter; we have as yet neither flowers nor leaves."*

In New England the Pliocene drifts are richly stanniferous, and occupying an extensive area, they will last for many years hence. Near Inverell, as indicated in my Report on that district, I have traced leads not yet touched by the miner.

The formations, therefore, of this period, present features of much scientific interest; and, especially in the Western Gold Fields, they are of great commercial value.

Lower Miocene.

In many parts of this Colony, as at Newstead, in New England, and near Gunning, in the southern district, at elevations of from 2,000 to 3,000 feet above sea level, occur deposits of pebble conglomerates, cement, clays, and ironstones, containing fossil

* "Fossils transmitted by the Rev. W. B. Clarke along with Rhytidocaryon Wilkinsonii—

Penteune Clarkei (variable in the size of its forms).

Phymatocaryon angulare (with a bivalved variety).

In reference to Spondylostrobus Smythii may be added that it produces (though very rarely) a *trivalved* variety.

F. v. M."

plants and leaves, similar to those from the leaf-beds of Bacchus Marsh, in Victoria. Of the latter, Professor M'Coy has observed that "the fossil plants of the ironstones are strikingly distinguished from the pliocene tertiary leaf-beds of the Daylesford and other older gold drift deposits, by the total absence of myrtaceous plants which so strongly mark the recent forest foliage of Victoria. I have no doubt the fossil leaves from this locality indicate a lower miocene or upper eocene tertiary flora, in which lauraceous plants form a remarkable feature. All the species seem new, but leaves of Laurus, Cinnamomum, Daphnogene, and possibly Acer, are scarcely to be distinguished from species referred to those genera in the leaf-beds (of the geological age mentioned) of Rott, near Bonn, and Oenningen (specially the Cinnamomum polymorphum, Heer.)"

In New England the Miocene drifts are in many parts very rich in stream tin. Those in the Inverell district will be found already described in my former Report on the Geology of that tin-bearing country. In the higher table-lands they also occur extensively. A very interesting account of the drifts or leads near Vegetable Creek has been given by the Mining Registrar, Mr. Geo. H. Gower, in his Official Report for 1874, from which I quote the following:—"Tin-mining in this district is only in its infancy, as, now that the deposits of tin have been nearly worked out of the bed of Vegetable Creek, the course of which for three and a half miles has proved very rich, the stanniferous wash has been traced in most of the mines into the banks and adjacent flats, which, in some instances, equals, if not excels in richness, the creek bed, and these discoveries have given increased confidence in the permanency and prosperity of the division.

"The lead on Messrs. Moore and Spears 160 acre purchased mineral land has been considered one of the best paying and richest mines—having declared dividend of £10,000 for the last twelve months' work (1874). On some parts of the lead the stripping (of a ferruginous sandstone and cement) has been very hard, requiring the use of blasting powder, but on an average all over the lead, which has proved payable to a width of 200 feet, the wash is easily got, and the depth of stanniferous dirt is 3 feet, averaging one bucket (80 lbs.) of tin ore to a load (sixty buckets.) The average weekly yield for the last six months, with fifty men and two horses, has been twelve tons. The amount of tin ore raised since the mine started (eighteen months ago) is 612 tons.

"Vegetable Creek Tin Mining Company," situated at a mile north of the line of main workings along the Vegetable Creek. A main tunnel has been driven 2,000 feet along the course of the lead, the width of which is from 18 feet, till it gradually widened out to 400 feet, where the present workings are, with an average thickness of 3 feet, of excellent paying dirt and at a depth of 60 feet from the surface. On one part of the workings, while sinking a shaft to cut the main lead, a very hard layer of cement 14 inches thick was struck, but under it a splendid run of wash was found from 2 to 4 feet thick, and 16 feet below this again the main lead is worked. But the sinking to the present is through all pipe-clay. The country at the 60 feet level is a regular river-bed, and in some parts there are fourteen feet of loose drift sand heavily intermixed with tin ore. Six hundred feet ahead of the Company's present workings, the ground tested by means of boring rods has proved that, after passing through 40 feet of extremely hard rock or cement, and under that a few feet of pipeclay, excellent, if not far richer, dirt is to be found, proving without a doubt that it is the richest and longest lead of stanniferous dirt ever struck at a depth in the Colonies. The yield of tin ore from the Vegetable Creek Co.'s mine for the past eighteen months is 350 tons, the average weekly yield is seven tons, with forty-five men and two horses."

"At Kangaroo Flat, Strathbogie Run, the tin is raised from a depth of 20 to 70 feet from surface. The sinking is through a volcanic basalt formation, and the alluvial deposit is a river-bed drift with occasional patches of cement, but with tin interspersed."

The drifts of the Miocene epoch, I believe, exceed all the other tertiary alluvial deposits in the permanency and productiveness of their yield of tin ore.

In taking a general view, therefore, of all the Tertiary or Cainozoic formations, including also the Recent, we cannot but perceive their high geologic and economic importance, yielding, as they will continue to do, the chief supply of the gold and tin production of New South Wales.

Mesozoic.

In the Northern District of the Colony, on the Clarence River, occurs a series of coal-bearing shales, conglomerates, and sandstones, containing fossil plants allied to Tæniopteris and Pecopteris (see specimens exhibited) supposed to be of mesozoic age. These beds have been described as being a long way above the Newcastle Glossopteris beds, and to have little in common with them. They belong to a different series, which may eventually be collated with the Mesozoic coal strata of Victoria.

Palæozoic.

The Palæozoic rocks have been so fully and accurately described by the Rev. W. B. Clarke in his "*Remarks on the Sedimentary Formations of New South Wales*" that it is needless for me to make further reference to them than is actually necessary in drawing attention to the geological collection now exhibited. Nearly the whole of the specimens from the coal measures were lately collected by myself, and they have been arranged according to their stratigraphical position.

The several divisions of the Upper Palæozoic series have been named after the localities where they are found to be most typically developed. In descending order they are as follows:—

Wianamatta series.
Hawkesbury series.
Upper coal measures or Newcastle, Wollongong, and Bowenfels series.
Upper marine beds.
Lower coal measures.
Lower marine beds.
Lepidodendron beds.

The Wianamatta beds consist chiefly of dark-coloured ferruginous, thin-bedded shales, containing fishes, shells, and plants. They have been much dislocated by faults or slides, as may be well observed in the numerous sections exposed in the railway cuttings between Sydney and Penrith. A splendid specimen of fish of the genus Palæoniscus is exhibited. It was found at the Gib Tunnel, on the Southern Railway, by Mr. Adams, the Surveyor General. With this is also exhibited another very perfect fossil fish, Cleithrolepis *granulatus*, of Palæozoic age, as determined by Sir P. de M. G. Egerton, Bart. This is the property of Mr. T. Brown, M.L.A., of Eskbank, and it was found in the Hawkesbury rocks on the Western railway line, at an elevation of nearly 3,000 feet above the sea. According to the Rev. Mr. Clarke, the cleithrolepis also occurs in the Wianamatta beds.

The Hawkesbury rocks are chiefly sandstones and grits of variegated colours, and upwards of 1,000 feet thick, near Sydney. At the head of the Lithgow Valley, near the Zig-zag, where they are well seen in section above the upper coal measures, they are over 400 feet thick, forming those bold, rocky escarpments and mural precipices which impart such grandeur to the scenery of the Blue Mountains. Sydney is built on the Hawkesbury sandstone, and many fine edifices of the city can testify as to the value of the stone for building purposes. Some of it, however, contains an objectionable quantity of salts. Where this is the case the stone weathers, not on the surfaces exposed to rain, but on the sheltered or under surfaces of the cornices of buildings. The same facts may be noticed in the sheltered hollows in the sandstone cliffs around Sydney Harbour, in the Blue Mountains, and especially in the alum caves in the sandstone cliffs near Wallerawang. The cause may be easily explained. As the moisture in the stone evaporates on coming to the surface, the salt held in solution crystallizes therefrom, and by the well-known power which the minute crystals exert when forming, they force asunder the grains of sand, and thus by degrees slowly disintegrate the rock ; whereas the upper surface of the stone is exposed to the rains, which continually wash away the salt as it appears, and so prevent its crystallization. The remedy should be either to coat the surface of the stone in dry weather with some material impervious to water ; or to reject stone that contains over a certain percentage of salt. A very simple process at once suggests itself for ascertaining the quantity of salt, by washing a certain weight of powdered stone, and then, after filtering the solution, obtaining from the latter the salt by evaporation. Such trials could be easily conducted, and would afford valuable information for architects and building contractors.

In the rocks in Sydney Harbour the Cleithrolepis granulatus has been found.

In many places the sandstone assumes a columnar structure similar to that so frequently to be observed in basalt. Some good specimens are exhibited showing the hexagonal columns ; these I obtained close to a dyke near M. La Perouse Monument at Botany Heads.

No coal seams of value occur either in the Wianamatta or Hawkesbury series.

Upper Coal Measures.

Almost all the fossil plants characteristic of the upper coal measures are exhibited, and were obtained from the beds at Newcastle, Bowenfels, and Wollongong. They include the abundant Glossopteris Browniana, Glossopteris linearis, Phyllotheca Australis, Phyllotheca Hookerii, Sphenopteris, Vertebraria, Conifers, seeds and stems of trees ; one dendrolite is a section of a silicified tree-stem, about 18 inches in diameter.

The splendid specimens I obtained from the Western Coal Fields are very perfect, amongst them may be seen several species of Glossopteris, a very large Vertebraria, and a peculiar peltate leaf, apparently new. With these I would draw attention to the fine impressions of an undescribed species of Conifer, which were found in the coal seam on the property of Mr. A. Brown, Coerwull, near Lithgow Valley ; also to a new fossil plant which I collected from the coal measures at Newcastle. Professor Dana, of America, who has examined the latter, pronounces it to be different from any plant he had yet seen. At Newcastle I have seen fragments of the frond of a Sphenopteris fern, which in its entire state must have measured five feet in length, and of Glossopteris leaves about two feet long.

In this series the Rev. Mr. Clarke, and Mr. J. Mackenzie, Examiner of Coal Fields, state that there are at least sixteen seams of coal, each more than 3 feet thick. The principal coal seam is from 8 to 10 feet thick; it is the same seam now worked in all the large mines of that locality.

Several faults are met with in most of the mines. One in particular I may instance, where the seam of coal has been bent in two places 90 feet apart, one end having gone up 35 feet and by lateral pressure thrust over the other a distance of 90 feet, causing the part of the seam between the two points of fracture to be overturned, the whole forming in section like the letter Z. The direction of the line of fault is about north and south. In various parts of the district trap dykes occur. In one mine where a basalt dyke, about 16 feet wide, cuts through the coal seam, the coal for 3 to 4 feet on either side of the dyke is much changed; the course of this dyke is north and south. In the collection exhibited will be seen a specimen of basalt attached to a piece of the coal seam through which it intruded.

On the coast near the Nobby, Newcastle, may be seen several trunks of trees up to one foot thick, with roots attached, starting from a seam of coal and embedded in the strata in the upright position in which they grew—a fact worthy of record, showing the comparatively quiet deposition of our coal beds as contrasted with the drifted appearance of the vegetable matter forming the irregular and thin seams of coal which I have observed in some of the Victorian beds. Mr. Selwyn has referred to the drifted origin of the material forming the coal as precluding the probability of the existence of workable coal seams in the carbonaceous series of Victoria.

The upper coal measures in the Western District are about 480 feet thick, resting conformably on the marine beds of the lower coal measures, and are overlaid by more than 500 feet of Hawkesbury sandstone. Eleven seams of coal have been counted in them; the lowest, which is 10 feet thick, lies about 25 feet above the marine beds, and is the same seam worked by the Bowenfels, Eskbank, Lithgrow Valley, and Vale of Clwydd collieries. Samples of coal from each of these mines are now exhibited. This seam of coal crops out at the surface on the railway line near Bowenfels. It dips at a low angle of from 3 to 5 degrees to the north-east, and is therefore easily mined, and as it passes under the vast extent of mountain ranges to the north and east it will be inexhaustible for generations to come.

The Hartley Vale Kerosene Co.'s Mine is situated about eight miles east from Bowenfels. The seam of kerosene shale or oil-bearing cannel coal is about 3 feet 2 inches in thickness, as shown in the following section, which I measured :—

BLUE SANDSTONE.

ft. in.
1 : 6 fireclay
1 clayband
3 black "casing"
4 "tops," or impure shale, yielding 40 gallons of crude oil per ton
6 to 8 black shale
½ „ 8 band of inferior fuller's earth
3 ferruginous black shale
½ to 8 wet pipeclay
3 : 2 kerosene shale, yielding up to 80 galls. refined oil per ton
10 "bottoms," yielding 60 galls. crude oil per ton
½ to 1 yellow band fuller's earth
bluish sandstone, with impressions of plants.

The Glossopteris is also found in the above kerosene shale.

The exact position of the kerosene shale I did not ascertain, but it is believed to be about 40 feet above the Lithgow Valley coal seam, which occurs here, only 2 feet thick; its thinning out being doubtless due to the proximity of the granite, which rises immediately to the south-west, thus cutting off the extension of the coal in that direction; but away from the granite towards the north-east, in the direction of the dip, there is every probability of the coal becoming thicker. The seam of kerosene shale crops out a little above the level of the river Lett, and it is easily worked by an adit driven into the face of the hill. The shale is sent by the Great Western Railway, a distance of eighty-three miles to the Company's works, near Sydney, where the "Comet Oil" for illuminating purposes, and other products, are manufactured from it; it is also exported for gas-works.

Similar oil shale occurs in the upper coal measures, as at America Creek, near Wollongong, where very extensive works have been erected for extracting the oil. At Berrima, and on the Upper Hunter, specimens of the shale from each of these localities are exhibited.

Referring to the kerosene shale seams, the late Mr. W. Keene, then (1867) Examiner of Coal Fields, remarks that "the discovery of our wealth in brown cannel oil coals and oil shales will enable us to manufacture all the oil needful for our own consumption, and even to export the raw article. We know that it exists in many places at wide areas apart, as may be seen by reference to the map; and like to the richness of our coal seams, which richness is not excelled in an equal vertical section in any part of the world, we may expect that the oil shales will be of as great importance in their development; and if we do not find oil springs, we may possess such beds of the solid material as will justify the expenditure of all the capital needful to keep up a steady and unfailing supply of the valuable and varied products which these shales and coals will yield.

"The works already in activity at Hartley and America Creek, and others preparing to operate in different localities, with the general approval of the quality of the oil produced, will justify our most sanguine anticipations on this subject."

A description of the collieries now working, and of the different varieties of coal raised, will be found in the last Annual Report of Mr. J. Mackenzie, F.G.S., Examiner of Coal Fields.

In the Bowenfels district there is an upper coal seam from 17 to 23 feet thick, but on account of the numerous clay bands it contains it is of less value than the lower 10-feet seam.

Lower Coal Measures.

The collection of fossils from near West Maitland, Greta, and Anvil Creek, includes Spiriferæ, Conulariæ, Inocerami, Productæ, Fenestellæ, Bellerophon, Crinoidal stems, &c., obtained from the upper marine beds 350 feet above the Anvil Creek coal seam, from which seam I collected the specimens now shown, containing the Phyllotheca and Glossopteris Browniana. Immediately below this coal are the lower marine beds, represented by the specimens of Spiriferæ, Conulariæ, Bellerophon, Pachydomus, Orthoceras, Euomphalus, Fenestellæ, a new species of Starfish, Chætetes radians, &c., from Stony Creek, Harper's Hill, Ravensfield and Singleton. This suite of fossils is specially interesting, as showing not only the range through the coal series of the Glossopteris and Phyllotheca, plants so abundant in the upper coal measures of Wollongong, Lithgow Valley, and Newcastle; but also, the association of those plants with the marine fauna of the lower coal measures; thus affording evidence agreeing with that of the previously mentioned fishes, as to the Upper Palæozoic age of the New South Wales coal measures.

The collection of Productæ, Spiriferæ, Euomphali (?), Conulariæ, &c., from the Bowenfels and Wallerawang District was obtained from coarse pebble conglomerate beds, about 100 feet below the ten feet coal seam of Lithgow Valley. These marine beds are conformable to the overlying plant beds of the upper coal measures, but rest unconformably on the upturned edges of Devonian strata.

Below the lower marine beds of the Hunter District are beds of shales and sandstones, with several species of Cyclopteris, Knorria, Sigilaria, Stigmaria, Lepidodendron, &c., (see specimens exhibited). They occur near to, and probably in association with, beds of marine fossils, which have been described as lower carboniferous. These fossiliferous strata, range from near Port Stephens, in the county of Gloucester, through the north-easterly parts of the counties of Durham and Brisbane, including the sources of the Hunter River; thence over the Dividing Range to Nundle and Goonoo Goonoo, on the Peel River, and again further north to the Manilla and Horton Rivers. The interesting section near Stroud, Port Stephens, which is now being made by the A.A. Company, under the direction of Mr. J. Mackenzie, the Examiner of Coal Fields, will afford much valuable information respecting the relative positions of the Lepidodendron beds and the Newcastle coal measures.

Devonian.

In the Rydal and Wallerawang Districts a considerable area is occupied by strata containing numerous Brachiopoda, and other fossils, which appears to be Devonian. The rocks consist of brecciated and pebble conglomerates, indurated quartzose, sandstones, and shales. They are traversed in places by thin quartz veins, which have doubtless contributed the gold which is always found in small quantities in the gullies intersecting these rocks. Amongst the characteristic fossils may be mentioned several species of Spiriferæ, which are very abundant, Rhynchonellæ, Orthis, Pecten, Murchisonia (?), Trilobites, Modiola (?), Corals, Crinites, &c.

The highest ranges in the district are of this formation. On and near Mount Lambie, which attains a height of 4,080 feet above sea level, and on Mount Walker, 3,803 feet, are beds of quartzose sandstone, about 1,300 feet thick, full of Spiriferæ. It is said that, from the summit of Mount Lambie, which has been cleared for the much-needed Trigonometrical Survey of the Colony, the late Sir Thomas Mitchell, and others, have seen the revolving light at Sydney Heads, a distance of about seventy-five miles.

Mount Walker lies six miles east from Mount Lambie; on the latter the beds have a general dip to the east, and on the former to the west, so that the intervening valley of the Cox River lies in a great synclinal trough of the Devonian strata.

About 1,000 feet above the Spirifera sandstones, are beds about 100 feet thick of hard quartzose sandstone, dipping east at 30°, containing the Lepidodendron which was first discovered near Mount Lambie House, by the Rev. Mr. Clarke and Mr. Francis Lord; five more specimens were afterwards obtained from the same place by myself and assistants, when engaged lately on the Geological Survey of that district. No fossils have yet been found in the sandstones and shales overlying the Lepidodendron beds; but I believe that they all, including the lower

Spirifera beds, belong to the same series, near the base of which are coarse pebble conglomerates, containing pebbles of limestone, enclosing Criuoid stems, Petraia, Favosites polymorpha, and other corals, probably silurian.

By actual measurement, I have ascertained the Devonian beds in the Rydal District, to be upwards of 10,000 feet thick.

From Mount Lambie, proceeding in a north-westerly direction, we pass from the indurated breccia conglomerates into lower beds of highly inclined shales and sandstones, with thick beds of Coraline limestone, probably upper silurian. Traversed by numerous quartz reefs, these rocks form the rough broken country to the westward, embracing the Mitchell's Creek diggings.

The geology, therefore, of this interesting district accords with the observations of the Rev. W. B. Clarke on the strata called by him *Devonian or "Passage beds,"* the fossils of which, he remarks, have at at once a silurian and a carboniferous aspect, being connected with the former by certain corals, and with the latter by the occurrence of Lepidodendron, Sigilaria, and other lower carboniferous plants.

"There is undoubtedly an apparent passage downwards from the marine fossils of the acknowledged lower carboniferous beds of New South Wales, to others which very much resemble the so-called Devonian beds of England, and a series of shells, corals, &c., from the Murrumbidgee, which I submitted some years ago to Messrs. Salter and Lonsdale, through Sir R. I. Murchison, Bart., excited doubts as to their belonging to any but Silurian and Carboniferous deposits. Among these were Phanerotinus, Loxonema, Atrypa *reticularis*, Orthis *resupinata*, Murchisonia, Strophomena, and Spirifera of various species, some like Devonian. Loxonema is known to me as occurring in the lower marine beds of the Hunter River basin.

"There appears to be an intermixture, and such is the case with certain strata to the westward of Wellington, in which some of the fossils have the Carboniferous type, and others the Silurian."

Referring to the lower Palæozoic rocks, the same author observes that "the greater mass of them appears to belong to Upper and Middle Silurian; the mudstones of Yarralumla, with Encrinurus and Calymene; the Coralline and Pentamerus beds of Delegat and Collalamine; the Tentaculite and Halysites beds of Wellington and Cavan; and the beds with Calymene, Encrinurus, Beyrichia, and others with Illænus, Harpes, Bronteus; Brachiopoda, including Strophodonta, and Radiata, embracing star-fishes, point to the existence of at least the Upper Silurian formation on both flanks of the southern part of the Cordillera. There are also numerous corals included in the list given by me in the Southern Gold Fields (p. 285), which also confirm the same determination;

and it may be added that the above and other fossils of this age mentioned by me elsewhere, have been examined by palæontologists of eminence in Europe. Such are the genera Favosites, Cænites, Ptychophyllum, Calamapora, Syriugopora, Emmonsia, Alveolites, Cystophyllum, &c. These perhaps might not alone satisfy a doubt, but with them occurs Receptaculites. Since 1858, when these were determined, I have detected Halysites, which may settle the question as to Upper Siluriau."

Specimens of Orthoceras, Encrinurus, Spirifera, Chætetes, Leptæua, Receptaculites Clarkii, Tentaculites, Halysites, Favosites, &c., from Wellington, Molong, Quedong, and Mudgee, are exhibited, together with Favosites, Gothlandica, Favosites, Polymorpha, Lathostrotion, Encrinital stems and Mollusks which I lately obtained from the limestone beds in the limestone reserve 7 miles from Wallerawang.

IGNEOUS.

The granites of New South Wales present an almost endless variety both in structure and in composition. In our collection are specimens from the northern, western, and southern districts of the Colony, of the ordinary ternary and porphyritic granites, fined grained eurite and coarsely crystalline binary granite, showing a very varied arrangement of their component minerals, quartz, felspar, and mica, the crystals of which may be seen from minute grains up to several inches in size. The granites are of various ages, but none appear to be younger than Carboniferous, unless the intrusive greenstone which has disturbed the upper coal measures near Scone, may be here included. Many of these are intrusive, while others are doubtless metamorphic, and still retain the bedding-planes of the transmuted sedimentary rocks; this latter feature is well shown in the banks of the creek near the township of Adelong. Several of the igneous rocks—granites of at least two ages, greenstone, and basalts of New England, are described in my report on the tin mines (published herewith). The Devonian strata of Mount Lambie have been upheaved and intruded by a porphyritic granite full of double hexagonal pyramids of quartz; it varies greatly in composition, and in places passes into a dense greenstone diorite.

Gold is found in the alluvial drift derived from these Devonian granites; and it is an interesting incident in the history of New South Wales, that in the granitic detritus near Hartley the Rev. Mr. Clarke first found gold in Australia, in the year 1841. That the *hornblendic* granites have always been found to be more or less auriferous was many years ago pointed out by the same distinguished geologist, whose valuable paper on the *Progress of Gold Discovery in Australia*, enters very fully into the subject.

In the Blue Mountains eruptions of trap have burst through the Upper Coal Measures and Hawkesbury sandstones, and overflowed them in places. The rich chocolate soil giving rise to such a luxuriant growth of vegetation on Mount Wilson has resulted from the surface decomposition of one of these trap dykes, which consists of hard dense angitic basalt containing crystals of glassy felspar (*oligoclase*). I believe this basalt to be of tertiary age, and its emission may have been probably contemporaneous with some of the extensive volcanic eruptions in the Western districts.

Gold.

Of alluvial gold two fine nuggets are shown from "M'Gniggan's Lead," about nine miles from Parkes. The larger nugget is pure gold, of dark colour, and weighs 22 ounces 18 dwts. 12 grs.; the other is scarcely less in size, but the gold is interspersed with quartz, indicating the existence of some promising reef, from which it originally came, in proximity to the place where it was discovered.

Rich auriferous quartz from the reefs in the Northern, Western, and Southern Gold Fields is exhibited. Amongst others are some fine specimens from the reef adjoining Foley's Folly, near Nundle, Peel River; from the reefs of Bingera, Tambaroora, and Oberon; from the "Sandstone Reef," Cowarbee, showing thin leaf-gold on the cleavage planes of the stone; and from the Adelong reefs, containing gold, galena, pyrites, and blende. On the Great Victoria Reef, Adelong, good stone is being raised from a depth of 525 feet. With the above are specimens from Major's Creek, near Braidwood, collected by the Warden, Mr. Charles de Boos; they are from a quartz reef two feet wide at a depth of 125 feet, and contain gold, sulphides of lead, silver, iron and antimony, and cobalt.

From Mitchell's Creek, county of Lincoln, Mr. J. Chiplin sent samples of pyritous quartz from a depth of 80 feet, the reef being 4 feet thick. Some thousands of tons of this quartz crushed, averaged 15 dwts. of gold per ton, and the pyrites yielded 10 ounces of gold per ton.

A specimen from Mr. J. Arkins, Mining Registrar, Cowra, contains coarse gold embedded in brown iron ore.

Quartz from Currajong, Billabong Gold Field, from Strickland's reef, eight miles north from Forbes, and from the Dayspring reef, with specimens of conglomerates, sandstones, slates, and limestone, illustrative of the gold rocks (Palæozoic) in the Lachlan District, have also been contributed by the Warden, Mr. Fred. Dalton, who in his annual report for last year says that one of the quartz reefs yielded from 100 tons, 1 ounce 7 dwts. of gold per ton. The Dayspring Company have crushed between the 1st of January and the 31st of December, 1874, 5,674 tons 13 cwt. of quartz raised from a mine 250 feet in depth, and taken from a lode 2 feet 6 inches in thickness invested by hard blue rock, and requiring to be removed by means of blasting powder. This stone produced 3,158 ounces 3 dwts. and 6 grains of gold, worth only £3 7s. 6d. per ounce; the average yield was 11 dwts. 12 grs. per ton. The hardness of the investing rock and the narrowness of the lode necessitates an unusually large expenditure in working this mine, the entire cost of raising and crushing the stone and extracting the gold is £1 11s. 2d. per ton; the value when obtained is £1 18s. 3d., leaving a net profit of 7s. 6d. per ton. Reefs abound in every direction that, if efficiently worked, will yield from 5 to 10 dwts. of gold per ton. The Day-spring is a fair representation of the reefs in the district.

Copper.

The following is a general description of the collection of copper ores exhibited.

Fifteen samples, collected by Mr. W. Bryant, from the Cow Flat Copper Mine, near Bathurst, consisting of gossan variegated ore (*erubescite*), green carbonate of copper (*malachite*), blue carbonate (azurite), and yellow sulphide of copper with galena. The samples were taken from near the surface and to a depth of 30 fathoms.

Armstrong Copper Mine, near Bathurst, sixteen specimens collected by Captain R. R. Armstrong from surface to depth of 45 feet—yellow sulphide, green and blue carbonates, red and black oxides of copper, yielding up to 35 per cent. copper. One specimen of green carbonate and red oxide of copper, with hydrous oxide of iron, gave on assay 35 per cent. copper, with 6 ozs. silver and 4 dwts. of gold per ton; another, principally yellow sulphide of copper, assayed 18·9 per cent. copper, with 3 ozs. silver and 7½ dwts. gold.

Wiseman's Creek, near Bathurst, from Mr. Cunningham; lode 18 feet thick, shaft 100 feet deep; blue and green carbonates, and yellow sulphide of copper, with galena; assay 19 per cent.

South Wiseman's Creek Copper mine, near Bathurst; lode 4 to 20 feet thick, strike N. 21° W., dip nearly vertical; 4 specimens from depths of 40 to 80 feet—yellow sulphide of copper with galena, yielding 28·8 per cent. copper, red oxide, green and blue carbonates, and grey and yellow sulphide of copper.

Great Western copper Mine, Milburn Creek, near Bathurst, six specimens consisting of red oxide, yellow sulphide, green and blue carbonates of copper. Also six samples of the same varieties of ore from the Milburn Creek Copper Mining Company, collected by Mr. J. Arkins, Mining Registrar.

From Apsley, south of Bathurst, Mr. E. Farr, Mining Registrar, forwarded samples of ore yielding from 20 to 30 per. cent. copper; they are of grey and yellow sulphide, sulphide of copper with galena, red and black oxides, and green and blue carbonates of copper, from depths of 20 to 90 feet, the lode being 3 to 5 feet wide.

Peelwood Copper Mine, ten miles from Tuena. From Mr. Thos. Taylor: green and blue carbonates of copper from twenty fathoms deep, red oxide and native copper from 30 fathoms, red oxide and green and blue carbonates from 40 fathoms, yellow sulphide of copper at 50 fathoms. The lode strikes S.E. and N.W., and averages 4 feet in thickness.

Ophir Copper Mine. From Messrs. Gilchrist and Weston; three large specimens of yellow sulphide and green and blue carbonates of copper.

Belara Copper Mine, near Mudgee. From Mr. C. R. Darton: samples of native or virgin copper, grey and yellow sulphide, green and blue carbonates, and red and black oxides of copper, assaying from 25 to 40·4 per cent.

Mr. F. Dalton, Warden, has sent specimens of yellow sulphide of copper from Buckinbah, and of green carbonates and red oxide, which he says abound north-west of Parkes, and south of Condobolin.

Messrs. Hurley and Wearne's Copper Mine, Wellington District: lode 2 feet thick, chiefly yellow and grey sulphide, with green carbonate of copper.

Gordon Brook, Clarence River. Bawden and Fisher's Copper Mine: good sample of coarsely crystalline yellow copper ore.

Frogshole, parish of Bala, county of King. From Mr. John Deer: green and blue carbonates, yellow sulphide, and red oxide of copper.

Bingera, Bobby Whitlow's Copper Mine. From Mining Registrar: red oxide, green and blue carbonates, and yellow sulphide of copper, from a north and south lode 2 to 3 feet wide.

Three-mile Flat, four miles north of Wellington. From Mr. J. Chiplin: green carbonate and grey sulphide of copper, from a lode 22 inches wide; and rich yellow sulphide of copper from a lode 18 inches wide, depth 88 feet, 10 miles north of Wellington. Mr. W. Edwards also sent samples of the above-mentioned ores, with black oxide of copper from the same localities.

Mitchell's Creek, county of Lincoln. From Mr. Gustavus Lett: Eleven samples, chiefly of yellow sulphide, with green and blue carbonates.

Wellington District. From Mr. W. B. Simpson: specimens of copper ore from nineteen different localities within 25 miles from Wellington; they consist of native copper, green and blue carbonates, and grey and yellow sulphide of copper—two specimens contain gold.

Cobar. From Messrs. Hardie and Gorman: beautifully crystallized specimens of *Azurite* and *Malachite.*

There are many other rich copper ore deposits not represented in the present collection, such as those in the Orange, Carcoar, Manero, Goulburn, Wilcannia, and Lower Murrumbidgee Districts. Nevertheless the above-mentioned localities alone afford evidence of the extensive distribution of copper ores in the Province of New South Wales.

In his Report for 1874 Mr. Whittingdale Johnstone, Warden of the Bathurst, Tambaroora, and Turon Mining District says: "I may here touch upon the mining for minerals other than gold. These are all copper mines, and are scattered over the Bathurst District, from Wiseman's Creek, near the upper waters of the Macquarie, to the vicinity of Cowra, on the Lachlan River. But I am not possessed of that statistical information or description in detail with regard to these mines which would enable me to report authoritatively upon their extent and productiveness. I may state, however, that although no great financial success has as yet attended the working of these mines, a large quantity of fine copper has been smelted on the spot; and that the existence of a rich belt of cupiferous country, only partially developed, has been conclusively proved to exist in the Bathurst Mining district, and with the advance of the railway, and the consequent increased facility for the carriage of ore and coal to and from the coal fields at Bowenfels, must eventually prove a material source of national wealth."

TIN.

The exhibits of tin ore from the Inverell District, New England, I have already alluded to. Those from the Vegetable Creek District, collected by Mr. Geo. H. Gower, Mining Registrar, are described by him in the following list:

No. 1. Two specimens. One is iron-stone (gossan), and the other is iron, shorl, and percentage of tin, from back of a lode found on the Glen Creek, five miles north of Vegetable Creek.

No. 2. Two specimens of quartz and tin ore, from the Glen Creek, five miles north of Vegetable Creek.

No. 3. Five (5) specimens of tin ore (shode stones), from the Grampian Hills, Strathbogie, a leasehold of 800 acres, the property of Messrs. Hall Bros. & Co., and situated 6 miles south-west of Vegetable Creek. These shode stones are found in the gullies at the Grampian Hills, varying from half-a-pound to forty pounds in weight. This property is a regular net-work of veins of quartz with tin crystals, from one-sixteenth of an inch to one inch wide; also veins of pure tin varying in width from one-sixteenth of an inch to one (1) inch. No defined lodes on this property have been opened, although there are indications of some extremely rich tin veins existing somewhere in the neighbourhood. The alluvial wash in the gullies about these hills is of a coarse gravelly description, and in some portions of this property the surface wash proves excellent. A sample of the stream tin obtained here I have sent (No. 19).

No. 4. Three (3) specimens of conglomerate (tin cement), from the Rose Valley Mine, adjoining south the Vegetable Creek Tin Mining Company (O'Daly's). This tin cement is found on top of the stanniferous wash of a river-bed drift, worked 30 feet from surface. The mine is not working at present, although it has turned out over 120 tons of ore since first operations were started. Two or three tribute parties have tried it lately, but could not make it pay on account of the very small layer of wash and about 18 inches to 3 feet of cement (as per sample), which would require a battery of stamps to operate on to extract the ore. The sinking is through pipeclay and sandstone, and the bottom is of a metamorphic granite formation.

No. 5. A specimen of tin ore from M'Master's Lode, situated at Tent Hill, 4 miles east of Vegetable Creek. M'Master's property consists of 260 acres. A large amount of stream tin has been obtained from the flats and gullies in it, all of which must have been fed by the two lodes which are traceable for a considerable distance through it.

One lode, which is traceable on the surface for a distance of 170 yards in length, and 2 feet wide, on which a shaft has been sunk 9 feet deep, and also a trench opened about 6 feet long and 2 feet deep; from both the shaft and trench large massive blocks of tin-stone, very rich in tin ore, have been thrown up. The other lode is situated on top of a high granite range, northerly from the one described above, and about thirty (30) chains distant. It shows on top of the range a large massive outcrop of quartz, traceable for about 15 chains in length. This lode is *six* feet wide, composed principally of quartz containing some rich stones of tin, a little arsenical pyrites, and a little copper (sulphide). On the whole surface of the range on both sides tin can be found. Judging from the strength of the lode, there is every indication to believe this is the champion lode of this district, although not as rich as the one mentioned above. The prospects of this property are such as to hold out every inducement to any enterprising capitalists for investment. The natural features of the country for the laying out of a dressing plant, and the supply of wood for driving purposes, are everything that can be desired. No more work has been done to test the value of these two lodes except that mentioned above.

No. 7. Two specimens of quartz and tin obtained from Ardern and Griffiths' Lode, Tent Hill, 2½ miles east of Vegetable Creek, at a depth of 12 feet from surface. At present no work is being done on this lode. A shaft is sunk to a depth of 35 feet, and the lode from surface has gradually improved, being now of a very defined character, 10 inches wide, and producing good stones of tin throughout.

No. 6. Two specimens from the same shaft as described above, in Ardern and Griffiths' property, but from 30 feet from surface. These two specimens are formed of quartz, tin ore, mispickel, sulphide of copper, with chalcedonic quartz cavities. The lode occurs in a clay-slate formation.

No. 8.—Specimens of quartz and tin from a prospecting shaft 6 feet deep, sunk on Elder & Co.'s ground, at the Graveyard Creek, one mile south of Vegetable Creek. No more work has been done.

No. 9.—Two specimens of faces of tin ore from the Great Britain Tin Mine, situated at the head of Vegetable Creek; stones like the samples are to be found all over the lead in this mine. The wash is of a stiffy clayey creek wash, heavily intermixed with large boulders, and lays on a clay-stone reef. In my report of 1st February last, on the Vegetable Creek Tin Mines, will be seen full particulars of this mine.

No. 10.—Specimens of petrified wood, from the Springs Strathbogie. The hills all around Kangaroo Flat and the Springs are covered with petrified wood, in fact on the leading range running parallel with the Glen Creek, a whole tree from the roots to the branches, lying on the ground, is petrified. The country is of a volcanic basalt formation, and tin ore is found at a depth of from 10 feet to 70, and 170 feet from surface. Messrs. Hall Bros. & Co., own large tracts of mineral ground in this part of the district, and have spent considerable sums of money in prospecting their ground, but have not yet, unfortunately, struck anything excessively rich.

No. 11.—A specimen of tin ore from M'Donald's property, Glen Creek, fourteen miles north of Vegetable Creek. This specimen is a sample of some coarse tin, which has been sluiced in a gully adjacent to a rich lode. The property is in the hands of Sydney speculators, but not any work has been done to give it a fair trial, yet it is recognized by all the old residents of this portion of the tin fields as being the richest and best show of a lode yet found in New England.

No. 12.—Two specimens of stanniferous wash-dirt from Ardern's 40-acre block, at the very head of Vegetable Creek. This block has been worked by several different parties of tributers, who worked out nearly all the creek bed, which is here a width of 6 to 10 yards; but lately another party, while prospecting the bank 30 feet from the bed, were fortunate enough to strike a rich vein of wash 2 feet thick running into the bank. The vein has been proved very good to a width of 30 feet, with every sign of its continuing; the stripping is sandstone and stiff clay, and the wash is a clayey creek-wash without very large stones, and requires puddling before the tin can be extracted.

No. 13.—Three specimens of stanniferous wash-dirt, from the Great Britain Tin Mine, situated west of Ardern's 40-acre block. (*Vide* No. 9.)

No. 14.—Sample of picked coarse pieces of stream tin from the Glen Creek, seven miles north of Vegetable Creek. (*Vide* No. 25.)

No. 15.—Sample of picked coarse pieces of stream tin from the Great Britain Tin Mine, Vegetable Creek. (*Vide* No. 9.)

No. 16.—A sample of stream tin from the Great Britain Tin Mine. (No. 9.)

No. 17.—A sample of stream tin from the Little Britain Mine, situated west of the Great Britain Mine, and about the centre of the principal operations in Vegetable Creek. This mine was considered worked out, as all the bed of the creek to a width of 120 feet had been wrought, but the energetic proprietor, Mr. L. R. Ashton, while prospecting a large flat alongside the old workings, struck rich wash one hundred yards away from the creek bed, and has since proved the whole flat, 400 feet square, to be payable. The stripping requires the use of blasting powder, having the appearance of hard ferruginous sandstone; the wash is 2 feet thick, of clayey creek-wash, with large stones, which requires puddling before operated on.

No. 18.—A sample of stream tin from Messrs. Hall Bros. & Co.'s Vegetable Creek Mine, situated at the lower end of the lead, worked in the bed of the creek. The lead has been proved payable in the banks, and is a hundred yards wide. The wash is heavy creek-wash with large boulders. Over several parts of the lead, from grass to bottom, is put through, but on an average the thickness of the wash-dirt is 2 feet. This is the only mine on Vegetable Creek where ground-sluicing is used, but then only when heavy rain falls. The apparatus for extracting the tin is the same as that used in most of the mines, *i.e.*, horse-pump, hopper-plate, and large boxes.

No. 19.—A sample of stream tin from the Grampian Hills, the property of Hall Bros. & Co., situated on Vegetable Creek, about six miles from the head of it. (No. 3.)

No. 20. A sample of stream tin from Tent Hill, Hall Bros. & Co.'s selection, 4 miles east of Vegetable Creek.

No. 21. A sample of stream tin from Kangaroo Flat, Strathbogie, 13 miles north-west of Vegetable Creek. In this part of the Division, tin ore can be seen on any of the hills by blowing the sand and gravel on surface. The gullies are not rich by any means, in fact no tin worth mentioning can be found in any one gulley or creek about here, they have such a rapid fall that the water runs off nearly dry within two days after a heavy thunderstorm. This only (the want of water) is the great drawback to Kangaroo Flat and its vicinity; also capital is wanted to give it a fair trial. The tin lies below a volcanic basalt formation, and the wash-dirt is a regular river-bed sandy drift, with good tin, and below this is a cement in which in some places the tin is very thick. This cement is very hard, in fact requires to be crushed by some stamping process to extract the tin. The bottom is composed of metamorphic granite. The sinking is from 10 feet to 70 and 90 feet deep, but the Hall Bros. have driven tunnels into the hills, having in some of them tramways laid. No work of any account in the way of tin-raising has been done this last six months, on account of the protracted drought.

In one portion of Messrs. Hall Bros. property here, stanniferous wash, *eight feet thick*, which would prospect from ¼ oz. to 1 oz. to the dish (tin prospecting dish) has been traced into the hill a distance of 100 feet. The proprietors deserve great credit for the enterprising manner in which they have prospected most of their numerous selections in this neighbourhood. That a very rich lead or numerous leads of rich stanniferous dirt exists in this vicinity there is every indication.

At the Sugarloaf Mountain, 3 miles west of Kangaroo Flat, and 16 from Vegetable Creek, also the property of Hall Bros., tin has been found, with very good prospects, 200 feet above the bed of the creek. It is one of the highest mountains for some distance around, and stands by itself in a sort of a hollow, with two gullies of good size on either side. The wash was followed into the hill by tunnels, but found to dip so suddenly as it were into the centre of the mountain, that many fancy it is a basin. The wash was very good as far as it was tested, a distance of 200 feet; but water having been struck, and the ground proving very treacherous, requiring a good deal of timbering, all further operations were abandoned for the present.

The bottom is metamorphic granite, and the stanniferous wash consists of round grains of tin ore (black) with double six-sided pyramids of quartz, also quartz pebbles, which are greatly water-worn, often found in the shape of hen's eggs, and a quantity of small pieces of white topaz and zircons in a decomposed white felspar (albite), almost in a state of China clay. [A small specimen forwarded, No. 27.]

No. 22. A sample of stream tin from "The Springs," Strathbogie. Hall Bros., & Co. own this property also. All the work is carried on here by adits driven into the mountain. The stanniferous wash lies under a volcanic basalt, at a depth of 170 feet from surface. It is situated 4 miles west of the Sugarloaf Mountain, Kangaroo Flat, and 20 miles north-west from Vegetable Creek.

No. 23. A sample of stream tin from the Gulf Stream Tin Mining Company, at the Gulf Creek, 20 miles north from Vegetable Creek.

It may be observed how very fine and gray this tin is, and the extraordinary fact of the ore in that part of the tin mines is that it is of the two extremes, either very fine and gray like this sample, or very heavy coarse lumps of tin the size of a pigeon's egg to 100 pounds.

No. 24. A sample of steam tin from Campbell Leasehold, situated a quarter mile south of the main workings in Vegetable Creek bed, and adjoins east of the Vegetable Creek Tin Mining Company's mine, and is on the same line of lead worked by the above Company, but in shallow ground.

Most of the dirt from grass to bed rock is sluiced, and pays handsomely.

No. 25. A sample of stream tin from the Glen Creek Tin Mining Company. This property is on the Glen Creek, and is seven miles north-east of Vegetable Creek. The surfacing on the hills in this portion of the Division is very good, but requires some different sort of washing process than what is used there at present (sluice forks and small boxes), for it is wash-dirt that does not require the extra expense of puddling; but a good portion of the large area of surfacing about the Glen Creek only requires some washing to despatch a large quantity of dirt through per day, for the prospects are not rich, but such a quantity of it could be broke at a small expense. Several lodes and quartz with tin have been found, and although large tracts of mineral land have been taken up for that description of mining, yet no work in the way of prospecting or testing the lodes has been done yet.

There are two large Companies own several hundred acres apiece on the Glen Creek—The Glen Creek Tin Mining Company, and The Banca Tin Mining Company—but no work of any account is being done on either of the properties.

The bottom or bed rock is clay-stone and also granite.

No. 26. A specimen of stream tin from the Glen Creek, at its junction with the Tent Hill Creek, four miles east of Vegetable Creek.

A great difference is observed between these two last samples (Nos. 25 and 26). Although they are both from the same creek, yet the tin is of a different colour—only to be accounted for by the different formation of the rock in the two places. Where this sample, No. 26, was got from, the bed rock is hard stratified granite.

No. 27. A specimen of wash-dirt from the Sugar-loaf Mountain, near Kangaroo Flat, Strathbogie, the property of Hall, Bros. and Company. (See No. 21.)

No. 28. A sample of stream tin from the deep lead worked in O'Daly's Claim (The Vegetable Creek Tin Mining Company), a full account of which I gave you in my report on the Vegetable Creek Mines, of the 1st February.)

No. 29. A sample of tin cement, as formed on top of the staniferous wash in O'Daly's claim, and in some parts of the lead is four feet thick.

No. 30. A specimen of wash-dirt from O'Daly's claim. This piece was knocked out of the *bottom* of wash-dirt, and gives you a good idea of how thick the tin lies in this lead.

Iron.

The iron ores exhibited are principally from the deposits near Wallerawang, discovered by Mr. H. Winters, and now the property of the Wallerawang Iron and Coal Company; from the Lithgow Valley Iron Mine at Eskbank, on the Western Railway line; and from Berrima, on the Great Southern Railway.

The Wallerawang Mines have been reported on by Professor Liversidge, of the Sydney University, whose analyses of the ores are fully given in his paper lately read before the Royal Society, and now republished. The ores consist of Magnetite averaging 40·89 per cent. of metallic iron; Garnet ironstone, containing 20 per cent. iron; Brown Hæmatite, 38·84 to 51·52 per cent. iron; clay bands, 49·28 to 56 per cent. iron. The magnetic ore sometimes associated with the garnet rock occurs in large irregular masses in altered Devonian sandstones near their junction with granite. The brown hæmatite forms a vertical lode in Devonian shales. The clay bands (brown hæmatite) of which five have been counted, average in thickness from 6 to 18 inches; they are inter-stratified with the coal measures immediately above the principal or lowest coal seam. Concretionary nodules of this ore lie scattered in considerable quantities on the sides of the rugged ranges where the clay bands outcrop. About two miles south-west from the coal mines and between it and the hæmatite and magnetic ore deposits, is a bed, about thirty feet thick, of brecciated marble limestone embedded with Devonian slates, so that within a radius of four miles Nature here presents the materials, iron-ore, coal, and limestone, for iron smelting.

At Mt. Lambie other out-crops of Magnetite with Micaceous iron occur in altered Devonian beds near the granite, as on the Wallerawang estate. The positions of these are all shown on my geological map of that district. The Lithgow Valley Iron Smelting Company propose to use the Mt. Lambie ore with their clay band ores, which latter (see specimens) are of the same character as those above mentioned, and occupy a similar position in the coal measures, which I have observed they maintain all through the district. The brown iron ore near the village of Bowenfels is from a mass cropping out on the surface for a distance of several chains.

A good specimen of Brown hematite is exhibited from the iron mines near Berrima. These mines were worked some years ago without success, this result being attributed to the unsuitable fue employed: but now that good smelting coal, it is said, can be obtained, we may hope that the failure will be only of a temporary nature.

Besides the above-mentioned iron deposits there occurs in the Illawarra district. on the property of the North Bulli Coal and Iron estate, a bed of argillaceous or clay iron ore, some twenty feet in thickness, in close proximity to the coal seams. It has been most favourable reported on by Mr. J. H. Thomas, C.E., Mr. T. W. Garlick, and others; and Mr. John Mackenzie, refering to it, says that, in all his experience of thirteen years actually employed in daily visiting the different mines in one of the largest coal and iron districts in Lancashire, England, he never saw a bed of clay iron ore so thick as this. Assays of the ore at the Sydney Branch of the Royal Mint gave respectively 32·9, 38·9, 55·7, and 44·3 per cent. iron.

The former Government Geologist, the late Mr. S. Stutchbury. mentions that, in the Great Park at Coombing, the property of Mr. Thomas Icely, there are five small hummocks (of from twenty to fifty acres each), composed of a very rich compact *hæmatite* iron, much of it being magnetic. A similar ore of iron forms the back of a copper lode on the same estate. The apparent quantity of iron is immense, and if all things else were compatible with the manufacture of iron, there is sufficient to supply another Sheffield for ages to come. The nearest coal, however, is that at Wallerawang, distant about seventy miles.

There are many other localities in New South Wales where rich iron ore abounds.

The reflection bears with it almost a reproach that, in this Colony, where there exists such immense and valuable deposits of iron ore, with suitable coal and limestone most favourably situated for smelting operations, the large and increasing demand for manufactured iron should have to be supplied from foreign sources.

Limestone—Colonial Marble.

The marble from Marulan and Cow Flat is exhibited in the form of tiles, which were cut and polished at the Woolloomooloo Steam Marble Works of Mr. John Young, who is now paving with this marble the floor of the great hall of the Sydney University. For such purposes it could not be excelled, as it is both hard and tough, and taking a good polish it would be useful also for mantelpieces and other plain decorative work. The Marulan marble is chiefly of a dark colour, sometimes black; that from Cow Flat varying from grey to white.

The specimens of marble which I lately obtained from the Limestone Reserve, near Wallerawang, are of several varieties of good colours. This marble dresses well, is capable of a high polish (as may be seen in the samples exhibited), and can be obtained in blocks of any required size and quantity. It is a very pure limestone, and as it occurs only about 7 miles from the Wallerawang Railway Station, distant from Sydney 105 miles, I believe that it will soon be a source of large supply for the metropolitan market.

The large polished slabs of marble from Tarrabandra, near Tumut, collected by Mr. E. A. Fitzgerald, are of rich colour, but the stone cannot be easily dressed on account of its "reedy" texture.

A polished specimen of fine "Sienna marble," from No. 6 Island, off the Queensland Coast, I have also placed in the collection. Should it occur in quantity, and be easily obtained, at no distant day there will be a ready demand for it in Sydney.

The beautifully crystallized specimen of *Calcite* exhibited, and the stalactitic limestone, were obtained from the Binda or Fish River Caves, by the Hon. John Lucas, Minister for Mines.

Miscellaneous.

Amongst the miscellaneous exhibits is some excellent *fireclay* shale, from the Lithgow Valley Iron Smelting Company's property, together with fire-bricks manufactured from it. Several such beds of fireclay occur throughout the upper coal measures of the Bowenfels and Wallerawang District. It will be of much value for the smelting works. I have also shown a sample of surface clay, from which the fire-bricks used in the furnaces of the Eskbank Copper Smelting Works are made.

Fluorspar, from Gow's Creek, near Wallerawang, is exhibited. It was lately discovered by me when engaged in the geological survey of that district, and occurs in small veins traversing a bed of felspathic rock 30 feet wide. Should it be found in greater quantity, it will be valuable as a flux for use in the reduction of copper ores.

The remarkably fine specimens of *Dendrites*—those moss-like forms of manganese oxide, often mistaken for vegetable impressions—are from the granite formation on Mr. J. Dobbie's property, Hilton, near Mount Lambie.

The collection of gem stones, from Mr. J. Hurley, of Mary Villa, comprises a large variety of both colonial and foreign gems. It includes many of the most valuable stones, and shows some of them both in their cut and uncut state.

In acknowledging the assistance received from officers of the Department of Mines, and others above mentioned, I would specially express my indebtedness to the Rev. W. B. Clarke, M.A., F.G.S., without whose published works but little would be known of the geology of New South Wales. We have, indeed, a few descriptive reports by other geologists, referring more particularly to certain localities, but the explorations of the Rev. Mr. Clarke have extended over not only nearly the whole of this province, but also through parts of the neighbouring Colonies; his researches are therefore of the greater scientific value, and should form the basis of future detailed geological surveys.

With the exception of the copper ores and some of the tin ores the greater part of the collection exhibited was lately obtained by the Geological Survey party under my direction. The Silurian specimens, however, are from the Museum of the Surveyor General, Mr. P. F. Adams, by whom and the Hon. J. S. Farnell, late Minister for Lands, the present Geological Survey was initiated. The necessity for this survey was long urged in promoting the interests of the Colony, which, from its almost unequalled mineral resources, appears destined to rank amongst the foremost mining countries of the world.

REMARKS ON THE SEDIMENTARY FORMATIONS OF NEW SOUTH WALES.

ILLUSTRATED BY REFERENCES TO OTHER PROVINCES OF AUSTRALASIA.

(By the Rev. W. B. CLARKE, M.A., F.G.S., F.R.G.S., Member of the Geological Societies of France and Austria, Vice-President of the Royal Society of New South Wales, &c., &c.)

3rd Edition.

[INTRODUCTORY NOTICE.—The *first* Edition of the following Memoir was written for and published in the "Catalogue of the Natural and Industrial Products of New South Wales," and forwarded to the Paris Universal Exhibition of 1867, by the New South Wales Exhibition Commissioners, at whose request it was undertaken.

It was re-printed at Melbourne in the "Official Record of the Intercolonial Exhibition of Australasia" in the same year, and was subsequently honoured by being transferred to the pages of the "American Journal of Science and Art."

The *second* Edition was prepared for the Report of the "Intercolonial Exhibition of 1870 at Sydney" and was included with an Essay "on the Progress of Gold Discovery in Australasia from 1860 to 1871" (by the same author), in the work entitled—"The Industrial Progress of New South Wales."

This, the *third* Edition, carries on the mention of geological experiences as to the Sedimentary Formations of Australasia to the present date, and has reference to the "Philadelphia International Exhibition of 1876"—of the New South Wales Commission for which, and of those of the years above named, the author has had the honor of being appointed a member.

Branthwaite, North Shore,
2nd June, 1875. W.B.C.]

If we inspect the map of Australia we observe that the coasts of Victoria, New South Wales, and Queensland, follow the general directions (with some irregularity) of the Cordillera, or elevated land separating the waters flowing directly to the coast from those which draining the interior, disembogue to the south-west.

The Murray River receives some parts of its tributaries from the high lands of Victoria, and others from New South Wales; whilst the Darling and its tributaries collect the remainder of the supply from as far north as 25° s.

The Cordillera thus sweeps round in an irregular curve from w. to e. to the head of the Murray—and thence, northerly and north-easterly, to the head of the Condamine; trending north-westerly from that point to 21° s., whence it strikes to the north, terminating its course at Cape Melville, in 14° s., about the meridian of 144° 30′ e., which is that of Mount Alexander in Victoria.

The more westerly and southerly trend of drainage is represented by the Thomson and Barcoo Rivers, which carry off the waters of the Cordillera at the back of the Barrier Ranges to Spencer's Gulf. The meridian of the head of that Gulf is, therefore, the western limit of East Australia.

The Cordillera itself, described in part by Strzelecki in 1845, was traced by him through a considerable part of its diversified course (as understood by him), from the southern point of Tasmania to the parallel of 28°, in longitude 152°; but not further westward than 146° on the parallel of Mount Alexander. It is, however, doubtful whether the range between this furthest western point and Wilson's Promontory, where he considers the chain to be cut off by the sea, forms anything more than a spur in that direction, though passing through Bass's Strait on to Tasmania.

But the extent of the Cordillera westerly, to its termination on the border of South Australia, is so well defined, that there can be no question that the s.w. and w. extension has as true a character as any part of the northern prolongation. This may be geologically deduced from researches of the Geological Survey of Victoria. That province is limited, at its eastern corner, by a line joining Cape Howe and the head of the Murray, so that the boundary crosses very near the highest point of all Australia, which Strzelecki made 6,500 feet above the sea, but which subsequent observations have shown to be 7,175 feet. This correction rests on observations made by myself in 1852, and on a re-discussion of them in comparison with results obtained by Professor Neumayer in 1862. On 8th May, 1852, I made the highest point of Kosciusco 4,077 feet above my then base, at 3,098 feet above the sea, which therefore came out 7,175 feet; and in February, 1863, Professor Neumayer wrote me word that he made the highest peak in November, 1832, 7,176 feet. This makes Koscinsco's summit, above the crossing place of the Indi or Hume River, at Groggan's, 5,425 feet.

To the northwards, the 144th meridian limits very nearly all the high land of the East Coast to Cape Melville, whilst the 142nd meridian limits to the westward the basin of the Darling, including part of the drainage along the Thomson and Barcoo, from the head of the Flinders to where it passes into South Australia on the 141st meridian.

Thus, all this enormous drainage of western New South Wales and south-western Queensland is, as it were, bounded by ranges of high geological antiquity, the Grey and Barrier groups being of undoubted similar age to the mass of the Eastern Cordillera.

It has long been known that the strike of the oldest Sedimentary rocks through the Cordillera, in Victoria, as well as in New South Wales, is generally meridional; so that in the former province the beds strike across the Cordillera, whilst in the latter they form various angles from parallelism with it to a transverse direction, as the chain doubles and winds irregularly in its course.

This is the experience of the Victoria survey, and my own traverses across various points of the Cordillera in New South Wales and Victoria establish the fact of a normal meridional strike of the oldest strata. So distinct, indeed, is this characteristic, that the settlers in various parts of this Colony have been accustomed to trace the direction of north and south by the strike of the slates, and are often guided by it.

It sometimes happens that, owing to the high angle of dip, and the effect of denudation on the overlying formations, the Cordillera itself becomes in places almost knife-edged, so that in New South Wales it presents occasionally a water-shed not more than nine paces in width; whilst in Maneero to the south, and in New England to the north, it spreads out in a plateau, on which eastern and western waters rise close together and sometimes overlap. These different features have a variable geological value as well as aspect; for, owing to the strike of the older rocks, the breadth of the Silurian formations, which, as in other countries, are repeated by recurring folds, may be more exposed in Victoria than it is in New South Wales; and owing to the curve of the Cordillera probably the same beds are traceable to the north which occur in the south; as, for example, the auriferous rocks of Omeo and Peak Downs, which are on the same meridian; and thus the meridional strike is exhibited along the north-east coast, where there are alternations of old rocks forming precipitous cliffs, with low valleys and beaches separating those alternations.

Independently of this arrangement the whole of the Central area inside the Eastern Cordillera has a trend to the south and west, so that the waters collected between 22° and 37° s., on the east of South Australia, find their way to the sea at the eastern corner of that province.

We might naturally assume that one order of deposits is to be expected throughout the Cordillera; but there is a singular exception. Whilst marine deposits of Tertiary age are found along the west coast of Australia, and along the southern coast from Cape Leeuwin to Cape Howe, there are no known *marine* Tertiaries in any part of the Coast of New South Wales and Queensland up to the Cape York Peninsula; and the reason of this may be, that, as indicated by phenomena before pointed out by me, but which on this occasion cannot be further dwelt upon, the eastern extension of Australia has been probably cut off by a general sinking, in accordance with the Barrier Reef theory of Mr. Darwin. This has some support from the fact that there is a repetition of Australian formations in the Louisiade Archipelago, New Caledonia, and New Zealand, in the latter of which occur abundant Tertiary deposits. The intervening ocean may, therefore, be supposed to cover either a great synclinal depression or a denuded series of folds; but, as shown in 1874 by the soundings from H.M.S. Challenger, this depression is of enormous depth, in one sounding 2,625 fathoms having been reached.

Relatively speaking then, the Cordillera of the eastern coast has not been subject to the changes which introduced the relics of a Tertiary ocean, or they have been removed by subsequent sinking and denudation. At any rate, no evidence is known to me of *marine* Tertiaries on the lands north of Cape Howe.

Another fact worthy of notice, as showing the probable ancient geological vicissitudes of Australia is, that the great Carboniferous series which is so prominent in New South Wales and in parts of Queensland, but which is less distributed in Victoria, and there only partially and irregularly as to the portions still remaining, has been broken up and carried away, so as to have left the various members dislocated, ruined, and separated in such a way as to allow no clear view to be taken of the whole till all the various portions have been separately examined; and to the want of this personal examination on the part of certain Palæontologists and others, who have never yet studied the Carboniferous formation of New South Wales, is to be attributed the perseverance with which they have so long disputed facts attested by geologists in New South Wales, who are familiar with that Colony and with Victoria also, but who are ignored by the closet-geologists of the latter.

In consequence of the absence of marine Tertiary deposits in New South Wales, and the occurrence of a more complete series of the strata in the sections of the Carboniferous formation, there has arisen a difficulty in collating the gold deposits with those of Victoria; and, in this respect, at present the upper deposits in the former province have not been assigned with much precision to the epochs adapted by Mr. Selwyn for the latter. And it also follows that his view of the distinct ages of Pliocene auriferous and Miocene non-auriferous gravels cannot be tested in New South Wales, if, indeed, it has not already been tested by the actual discovery of gold in the so-called Miocene deposits themselves, as they occur in Victoria.

So far as is at present known, gold in Victoria is derived chiefly from the Lower Silurian formation; but researches conducted for me at H.M. Mint in Sydney prove that it exists in almost every distinctive rock of New South Wales. In this province the alluvial deposits are not so extensive as in Victoria; but this probably arises from the fact previously mentioned of the strike being in Victoria transverse to the direction of the Cordillera; by which means the currents which distributed the drift had a wider area of gold-bearing materials to denude than in New South Wales, where, I conclude from numerous examples, the principal currents were to northward, so that in that province they would coincide with the direction of the Cordillera, and not accumulate the deposits in such low-lying extensive regions as those of the Murray Districts. The same objection would obtain on the supposition of gradual waste and accumulation from less powerful agency than that of a general rush of water. It is not, however, to be doubted that there is an enormous amount of gold yet untouched in numerous places in New South Wales, not only in the quartz lodes (or reefs) but in gullies and plains where alluvial gold diggings will yet be discovered.

Dr. Duncan, in an elaborate paper on some of the fossil Tertiary corals of Australia (*Proceedings of the Geological Society, August*, 1870), suggests the propriety of discarding the divisions into Pliocene, Miocene, and Eocene, of the Australian Tertiaries, and of substituting the general term Kainozoic, since he considers them merely as successive deposits of one continuous epoch. But, as proved by my own researches more than twenty years ago, much of the gold in New South Wales is derived from iron pyrites in granite, and in *beds* of sedimentary origin, consisting of siliceous matter cemented by iron derived from decomposed pyrites, whilst it has been shown by Aplin, Daintree, Hacket, Wilkinson, and others, that much gold in Victoria and Queensland is due to the intrusive agency of felstones, elvanites, and diorite. The dykes or reefs of quartz in the Silurians are therefore not, as once supposed, the exclusive sources of Australian gold. Nay, there is good reason to believe that the Carboniferous rocks are themselves impregnated, as in one remarkable instance on Peak Downs. In New Zealand gold sometimes occurs so mixed with siliceous particles as to constitute with them a golden sandstone.

The distinctive differences in material mineral wealth between Victoria and New South Wales are not altogether confined to gold or tin, which latter metal is well represented in New South Wales and Queensland; but coal, iron, and copper, and perhaps lead, prove together more than an equivalent of the great amount of gold in Victoria.

At the Universal Exhibition of 1854-5, the present writer exhibited a collection of rocks and fossils, illustrating the whole of the geological formations of Australia as then known, and these were enumerated in their stratigraphical order in the published Catalogue. A few remarks on the various geological epochs, as they now represent themselves in New South Wales, with brief statements as to their connection with other portions of Australasia, may be all that is necessary on the present occasion.

Azoic and "Metamorphic" Rocks.

There has not been sufficient evidence yet collected to show that these rocks extensively exist in Eastern Australia, although in Tasmania rocks of a doubtful class (and which may, perhaps, be only highly altered Lower Silurian) have been referred to them by Mr. Gould. The existence of gneissoid strata, and of schists of very ancient aspect, are also well known in New South Wales, with occasional unfossiliferous limestones; but it would be premature to place them, without doubt, under the present head. Mr. Daintree, however, describes them as the source of some gold in the Cape River and Gilbert Districts, to the north. Some of those mentioned under the "First Epoch" of Strzelecki have, on close inspection, appeared to me to be merely the products of transmutation; nor is such an improbable result, seeing that in Australia some slates have apparently been changed into granitic rocks. It is at least certain that such rocks generally occur in the immediate vicinity of granites, which latter frequently occupy large areas both in Maneero and in New England, as well as along the Cordillera, and in independent masses along the coast. In Western Australia, where an enormous region is occupied by granites, and the older formations are represented only by small patches of slates, whilst the granites themselves remain bare, these patches are found on the flanks of the granitic bosses and at extremely wide intervals; nor have I been able to detect among the numerous collections which have passed through my hands, any distinct evidence of any but doubtful examples of those foliated rocks which belong to the so-called primary epoch. In

Southern Australia, also, there does not appear to be any considerable amount of strata which could be referred to this epoch. Transmutation has, however, acted vigorously in New South Wales.

Lower Palæozoic Rocks.

(*Lower and Upper Silurian.*)

Of these there are undoubted evidences in some limited districts of Tasmania and Queensland, whilst in Victoria and New South Wales considerable areas are occupied by them.

Western Australia has as yet not furnished any fossils of Silurian age; but, according to Mr. Y. L. Brown, Government Geologist, there are clay slates, schists, and other rocks which may be Silurian much transmuted, judging from their position and composition.

The North-west territory is in much the same condition.

South Australia has furnished two fossils, *Pentamerus oblongus* and *Cruziana cucurbita*, stated by the Rev. Julian E. T. Woods, in his account of the Geology of that Colony (p. 20 and 21), as belonging to the Silurian epoch. The former occurs in New South Wales; the latter in the Bolivian Andes.

In Tasmania along the Gordon and Franklin Rivers occur various Silurian fossils, some among which identical with those of New South Wales were noticed by me; but Mr. Gould considers others to be Lower Silurian. This formation evidently exists in that Colony, for in 1873 I received from Mr. T. Stevens, F.G.S., some Trilobite-sandstone from the western part of the Island, which Mr. Etheridge determined for me to contain *Phacops*, *Ogygia* and *Calymene*; and to these Professor Bradley, of the U.S., to whom was forwarded by me some of the rock, has added *Conocephalites*, thus proving the relations of the rock to the Potsdam sandstone.

Mr. Gould mentioned, in June, 1860, a *Calymene* at the base of the Eldon Range. I found that genus also in New South Wales in 1852. In Victoria Professor M'Coy has made a list of twenty-five Lower and fifty-three Upper Silurian fossils, including in the former twenty-three Hydroid zoophytes, and another species belonging to the Upper formation. Of the Graptolitidæ only one is said to have been found in this Colony, and I presume that it is more likely to belong to the Upper Silurian than to the Lower, though towards the Victorian boundary, along the Deleget River, Lower Silurian rocks, according to some, are supposed to make their appearance.

New South Wales offers a more determined evidence of the existence of certain Silurian deposits, but singularly enough nothing has been positively shown of the existence of any fossils below the base of the Llandovery or the Middle Silurian, except in the case just mentioned.

To this epoch I referred fossils found by me in Maneero, in my Report of November, 1851, which was re-published in 1860, and it is satisfactory to find that the examination of a considerable amount of specimens by Prof. de Koninck of Liége, who kindly undertook the task of describing them, has resulted in a confirmation of my opinion.

Summing up his review of sixty of these, he says that they are in nearly equal divisions of the upper and lower beds of the Upper Silurian formation, and that they closely agree with the fossils of Europe and America; that the major portion of the former belongs to the Actinozarians and Crustaceans; and that the latter are nearly all Mollusca; and that none of the Graptolites noticed by Prof. M'Coy in 1861, and more recently by Mr. R. Etheridge, junr., from the Victorian strata, occur in the collection sent by me. And he concludes, as I have done, that at present the existence of fossil beds below the Middle Silurian has not yet been determined in New South Wales.

It is otherwise in Victoria, but it may be that some of the highly transmuted rocks of the south-west portion of New South Wales may yet furnish traces of greater antiquity when thoroughly examined. In the last edition of this Memoir, published in 1870, I mentioned the existence of certain Corals, Trilobites, &c., as determined for me in 1858 by the late Messrs. Salter and Lonsdale.

Professor de Koninck is not in antagonism with those geologists, but in the fresh series of my fossils he found among the trilobites Staurocephalus, Cromus, Proetus and Lichas, in addition to Calymene, Encrinurus, Illænus, Harpes, and Bronteus before announced by myself. (See edition of 1870, p. 6., and Southern Gold Fields, 1860, p. 286.)

In due time I hope to publish all the data connected with these and other associated fossils of the Upper, Middle, and Lower Palæzoic formations.

Nothing lower than Siluro-Devonian, according to Mr. Etheridge (in reviewof Mr. Daintree's fossils, Q.J.G.S., August, 1872), had up to that time been found in Queensland. But as elsewhere mentioned, I considered the Brisbane slates to be analagous with those of the Anderson Creek Gold Field in Victoria, both of which groups I examined personally *in situ.* The latter are held to be Upper Silurian.

I am inclined to think that there may yet be found in some of the deep gullies and ravines, outcrops of the lower rocks which have escaped notice. But the *fossil* evidence is, at present, not confirmatory of that opinion.

Middle Palæozoic Rocks.

The late Mr. Jukes desired the term Devonian to be eliminated, referring the so-called beds to the bottom of the Carboniferous formation; but geologists have not generally accepted that proposal. The series of shells, corals, &c., from the Murrumbidgee, which I submitted in 1858 to Messrs. Salter and Lonsdale, through Sir R. I. Murchison, Bart.,* excited doubts as to their belonging to any but Silurian and Carboniferous deposits. Among these were Phanerotinus, Loxonema, Atrypa *reticularis*, Orthis *resupinata*, Murchisonia, Strophomena, and Spirifera of various species.

Mr. Salter's Report to me was as follows: "These fossils are of a mixed character, many being of unquestionable Silurian age, and others having all the aspect of Carboniferous and Devonian fossils. It will not be so easy to predicate those of Devonian type, as there is much similarity between fossils of that age and those of either of the other systems, the Lower Devonian species being very like Silurian, and the Upper like Carboniferous ones. But if none of the fossils came from Carboniferous beds, then there must certainly be Devonian forms mixed with Upper Silurian."

Mr. Morris contributed, in 1845, a paper to Strzelecki's work of that year, in which he says: "The Palæozoic series of Australia and Tasmania may be regarded as partly the equivalent of the Devonian and Carboniferous systems of other countries."

In 1861 (*Cat. Vict. Exh.*) Professor M'Coy stated that "there had as yet been no exact identifications to prove the existence in Australia of the intermediate Middle Palæozoic or Devonian formation." And as recently as 1866, Vicomte d'Archiac (*Géologie et Paléontologie*, p. 468), writes thus: "Le développement des séries siluriennes et carbonifères dans l'Australie doit y faire soupçonner entre elles un représentant de celle qui vient de nous occuper; mais il ne semble pas qu'elle y ait encore été bien charactérisée par ses fossiles."

About the same time Professor M'Coy (Exhibition Essays of 1866-7) mentioned that the limestones of Buchan, in Gippsland, contained "characteristic corals, *Placodermatous* fish and abundance of *Spirifera lævicostata*, perfectly identical with specimens from the European Devonian limestones of the Eifel." In the Official Record of the Exhibitions of 1872-3, the addition of some other places in Gippsland (unnamed) and of Mount Gibbo, is introduced by the Under Secretary of Mines for Victoria; and in 1874 also, Mr. R. Brough Smyth included in his "Progress Report of the Geological Survey of Victoria," a list of fossils of

* See Murchison's "Siluria," 3d ed., p. 296, and 4th. ed., p. 276 and p. 462.

the most characteristic common types, drawn up by Professor M'Coy, which, under the head of Devonian, includes the following: *Favosites* (two species), *Spirifera lævicostata, Grammysia* (n. sp.), *Orthonota* (n. s.), *Asterolepis* (plates allied to). In 1847 the same skilful Palæontologist noticed some striking resemblances to Devonian fossils in a few of the large collection I sent in to the Woodwardian Museum at Cambridge; and Professor de Koninck, also in 1847 (*Recherches sur animaux fossiles*) records Sp. *Murchisonianus*, a Devonian fossil from Tasmania.

In order to test the existence of a wide-spread Devonian series in New South Wales, I requested (as stated elsewhere) my friend Professor de Koninck, to undertake the examination of a collection of 1,000 Palæozoic fossils, comprising the Upper, Middle, and Lower Palæozoic formations as they exist here, and he has just favoured me with his account of the Devonian forms, concluding it as follows:—

"Of 81 species observed, there are but five belonging to the Upper Devonian, all the rest are of lower beds. Of these 81, thirty are new to science, and are Australian; but save four, all have their types in Europe and America, and have the same character and position as those."

Amongst these the Professor includes the fossils I referred to in the last edition of this Memoir (p. 10), from Yass, Mount Lambie, and on the Turon and Moruya Rivers, and which are in part, identical with the Mount Wyatt shells in Queensland. These latter are mostly Brachiopods, and I have collected them during my different journeys of several years from the western boundary of the Carboniferous formation (underlying it *in situ*), and occasionally from a scattered over-lying drift, ranging for nearly 260 miles of direct distance (included between 36° south on the Moruya, to nearly 32° south.) The principal of these particular Brachiopods, are:—*Rhynconella pleurodon, R. pugnus, Spirifer disjunctus, S. Yassensis; Orthidæ, Productæ, &c.* They occur *in situ* between the slaty rocks of Sofala and the overlying Carboniferous beds on the Turon; south of Moruya River; near Mullamuddy on the Cudgegong River; at Cudgegong Creek; in the deep defiles of the Upper Colo River; and in other places. Mr. C. S. Wilkinson, with whom I visited the locality a month or two ago, found them under interesting circumstances occurring in a great synclinal curve, from nearly the summits of Mount Lambie and Mount Walker (with considerable dips), and explaining the sources from which the loose pebbles collected by me at Bowenfells some years since, were probably derived. From the occurrence of different fossils in the pebbles, it is certain that many strata of the Devonian formation must have been broken up, and it seems that similar beds have undergone the same process in other countries, for I well remember picking up, in 1829, in the "Platz" of Coblentz on the Rhine, a similar drift pebble, of just such rock as that in question, containing a Brachiopod of like age.

During some recent explorations in the north-west of this Colony, I became satisfied as to the widely-spread extent of the Devonian series, of which more evidence will be elicited hereafter, the data for which are already sufficient, but there is no room to introduce them on this occasion.

I may add here, that De Koninck considered the fossils he examined to be above the European strata with *Calceola;* but though not present therewith, Calceola occurs at Mount Frome, in the county of Phillip, and Streptorhyncus elsewhere.

Tasmania gives no well-established proof of the existence of Devonian rock. But it is a fair inference, first suggested by the late Mr. Salter, that the broad-winged Spirifers common there in the Palæozoic beds imply the probable occurrence. Mr. Jukes and Mr. Gould both repeated the inference. Mr. Darwin and Mr. Selwyn agree that some of the Tasmanian fossils "occur in the Silurian, Devonian, and Carboniferous strata of Europe." This is nearly all that is known respecting their position.

Western Australia, according to Mr. Brown's Report, adds nothing to the history of the Middle Palæozoics; but Mr. H. Gregory indicated on his map and in his Report the existence of Devonian rocks near York, and in other parts of that Colony. Having examined the rocks so indicated, I can only state my belief that they have no pretension to any such antiquity, and are probably mere collections of loose granitic matter and other drift cemented by ferruginous paste, which has since become transmuted into concretionary nodules and hæmatite. There are also pebbles of trap, much decomposed, in the so-called Devonian. They may perhaps be more properly considered as representing the *laterite* of India.

Queensland, on the other hand, exhibits a stretch of Devonians extending through *ten* degrees of latitude. Not the least interesting facts are that the Tin Mines of Queensland (as well as those of New South Wales) occur in granites of Devonian age.

At Gympie, on the river Mary, rich gold-bearing quartz reefs occur in transmuted slates and other tilted beds, which are composed of detrited dioritic matter and brecciated deposits, in which are abundance of fossils of doubtful aspect, and these I before referred to some part of the Carboniferous formation. Mr. Etheridge considers and has described the fossils as Devonian. They certainly have much in common with the Devonian beds of North Germany and Belgium, described by Sedgwick and Murchison, as I stated in the Second edition, p. 10. It is right, however, to remark that Professor M'Coy does not adopt this determination, considering the rocks to be younger.

In rocks of the same age also a vast deal of mineral wealth of other kinds occurs, as ores of copper, iron, lead, antimony, &c.

In the notes on the Geology of Queensland, by Mr. Daintree, (Q.J.G.S., Aug., 1872), the fossils are described and figured by Mr. Etheridge; and to that excellent Memoir the reader is referred for much valuable information. Twelve species of the fossils are described as Devonian.

It is interesting to find Dr. Hector stating at the beginning of 1875, that 2,000 specimens of Lower Devonian or Upper Silurian fossils have been obtained from the north-west district of the South island of New Zealand (*Ninth Annual Report of the Colonial Museum*, 1874.) And equally interesting is it to know that New Caledonia also holds out hope of contribution to the Middle and Lower Palæozoic faunas, as in the Isle Ducos, Leptæna, Spirifera, Orthis, &c., occur with *rolled Brachiopods* of the same character as those at the "Gulf" on the Turon River in this Colony. (*Annales des Mines, tome* XII, p. 54, 1867.) Monsieur M. P. Fischer is disposed to assign them to the *Devonian* period (*Bulletin de la Soc. Géol. de France*, 18 *Mar.*, 1867.)

It may be well to mention that the *Old Red* exhibits itself in association with the limestone and slaty portions of the formation, occupying ranges of considerable extent and prominent character in the Western districts, and that a Lepidodendron of some local interest (*L. nothum*) also occurs in three of the Colonies.

It seems as if every individual discovery in the Geology of this Colony had a history or literature of its own.

In June, 1851, Professor M'Coy wrote to me from Cambridge respecting the first Lepidodendron he had seen from Australia, and which I had forwarded by the late Rear-Admiral King to Professor Sedgwick, and stated it to be *L. tetragonum* of the English coal fields.

The late Mr. Salter, in his letter to me of May 9, 1856, said, however, that the genus was *not Lepidodendron*.

In November, 1863, Sir C. Bunbury wrote to Professor R. Jones, respecting a collection of Australian fossil plants including the above species sent home by me, and now in the Museum of the Geological Society, where they were inspected by him, at my request, and noticed one (*the* one) which he considered to be very like *L. tetragonum*.

During the last few years I have collected, or received, this plant from a variety of localities on New South Wales and Queensland, and from the latter Colony it was also brought in abundance by Mr. Daintree. Mr. Carruthers, who has given its description fully in the paper before alluded to (Q.J.G.S., Aug., 1872) has assigned to it the name of a species described by Unger, viz., *Lepidodendron nothum*.

The extent of territory from which my specimens have been collected embraces a direct distance of more than 1,100 miles (English) between 19° S. and 35° S. (of course at intervals only), from which we may infer the importance of its discovery in any new locality, as establishing the existence of a portion of the Devonian series to which it has been finally assigned.

Till the present date it was not surprising that even careful observers should classify this plant with Lower Carboniferous species, as Mr. Odernheimer did in his paper on the Peel River Estate (*Sydney Exhib. Catalogue*, 1854, p. 54), and as I was *reproved* for doing in 1851 (Report on Coal Fields, Western Port, Victoria, 1872.) If M'Coy was *right* in that instance, I could not be far *wrong*.

It was satisfactory to be able to recognize this plant in January last in a creek near Rydal, on a spur of the Mount Lambie Range, where the Devonian Brachiopoda occur, and to be able to direct Mr. Wilkinson to the locality where he found his five additional specimens, which certainly establish the position *in situ* of the species in that locality.

The quotation from the Coal Report named above, and the assertion of the Reporter, show that the opinion held by Professor M'Coy as to the age of the Lepidodendron in question is still maintained.

In the first Decade of his excellent work, illustrating the Palæontology of Victoria, now in the course of publication, he combats in a moderate tone the assignment of this plant to *L. nothum*, still re-asserting his old opinion.

Writing in 1861 the learned Professor proves that there is no mistake about the identity of the plant in question, for he says, a specimen of it, still I believe in the Melbourne Museum, is of the same species as *the only Palæozoic coal plant* ever collected in New South Wales, and which was sent to me about twelve years ago for "determination during the controversy as to the age of the plant beds of the Newcastle N.S.W. beds." This mistake as to *date* is of no importance, as it is rectified by my previous quotation from Mr. M'Coy's letter, and I only refer to it to show, which is due to himself, that we are treating of one and the same plant.

Upper Palæozoic.

I would not venture to say, that no Lepidodendroid plant is to be found in our coal measures, or even the one in question, if the range of that species goes upward; for I myself submitted two coal plants from Lower Carboniferous rocks on the Rouchel River to Professor Dana, who sent them on to Mr. Leo Lesquereux of Columbus, Ohio, the best authority in America, on fossil botany, and whose report is that one is near *Lepidodendron dichotomum* and the other is *L. rimosum* of Sternberg, and that both are undoubted plants of the true (European and American) coal measures.

Having personally compared with specimens from Kiltorkan (in my possession) the *Syringodendron dichotomum* (of Mr. Carruthers' paper before referred to) which I sent home to England some years since, and which is yet in the Geological Society's Museum, let me add that I found it in company with the *Lepidodendron nothum* and some other casts of plants, in the year 1852.

I would remark, that in one locality in Tasmania I collected many individuals of a species of so called Syringodendron, which occurred in the coal measures at the base of Spring Hill, on the slope of which hill Strzelecki stated that he found in beds of sandstone *Pecopteris odontopteroides* underlying the *Pachydomus globosus*, known to Professor M'Coy as a Wollongong Lower Carboniferous shell. It is only fair to add that though I made in two different years a close examination of the hill and the surrounding district I failed to recognise the shell, though I saw much that reminded me of the geology of certain parts of the Hunter River coal formation, and of the Illawarra, of the age of which there is little doubt.

On the borders of the Devonian formation in parts of the Hunter and Manning River basins, the Lower Carboniferous which is highly inclined passes on along the same strike into beds charged with *Lepidodendron, Knorria, Sigillaria*, &c., and in some instances Lepidodendron occurs in the same blocks with? *Otopteris ovata* of M'Coy, an example of which was shown in the Exhibition at Sydney in April, 1875, from the east of Stroud. On the ranges at the head of the Peel, and about Booral, Stroud, and Scone occur numerous fragmentary blocks with Lepidodendron and other usually associated fossils of what by many would be considered Lower Carboniferous beds.

These and other facts of similar kind have been often stated by me on former occasions. They are referred to on this, in order to show the relations of the New South Wales formations. At present many of the points where the Upper and Middle Palæozoics meet are ill-defined, and it will require the researches and labours of many years to fill them in with strict accuracy. Nor can it be wondered at, that in so large a territory and with such complicated and broken features details must for a long period to come give way to generalizations. Aware of what is wanting I, nevertheless, accept with satisfaction the testimony offered to the work I have endeavoured to perform, because what has been accomplished by me single-handed, and without the aid which workers in such a field expect and receive from public funds, has been undertaken and carried out, so far as has been practicable, with singleness of purpose and in reliance on my own resources.

In the course of my work I have tried to contend with the prejudices of some who *have never visited this territory*, and who, from a distance of many hundred miles, have ventured to dogmatise, solely from a palæontological point of view, without caring to ascertain how far the stratigraphical evidence is at variance with their conclusions.

In consequence of this the ascending order of formations above the Lower Carboniferous in this Colony has long been disputed by some, whose unacquaintance with facts, patent to all who have examined them, is the best apology for a more temperate style of criticism than has been adopted.

We are indebted to Professor M'Coy, for ascertaining, in 1847, the existence of eighty-three species of animal remains in our Carboniferous formation, in a collection forwarded by me to the University of Cambridge, in which the Professor was then officially employed.

Before that time, Bowerbank, Sowerby, Morris, and Dana had determined the existence of the Carboniferous marine beds; and the latter author enumerates about eighty species observed during his excursions in New South Wales, in some of which I accompanied him.

More recently Mr. Etheridge has described fifteen species of Lower Carboniferous fossils from Queensland, in relation to Mr. Daintree's paper on the geology of that Colony, of which ten were furnished by myself. None have yet been discovered in Victoria. In Tasmania, Mr. Gould figured some well known forms from that Colony, but the plates were never published.

He has noticed also what I have contended for, that the worked coal beds of the Mersey River belong to the same formation with Palæozoic marine fossils, as in Queensland and on the Hunter River.

Having visited the Tasmanian locality for the purpose of inspection, I can confirm all that has been stated respecting the occurrence of the Palæozoic fossils, Orthonota, Spirifera, Fenestella, Pachydomus, Theca, &c., in association with and immediately above the coal; and within the last few months I have been officially informed that coal seams have been found by piercing these beds on the Don River, confirming my grounds for recommendation to look for them.

In Western Australia traces of these marine beds have been detected and announced by Mr. Gregory. And in extension of the formation northwards beyond the limits of Australia, it is well known by more than one observer, that Carboniferous beds exist in the island of Timor, where Beyrick discovered several of our New South Wales species, *e.g.*, *Spirifer lineatus*, *Sp. Tasmaniensis*, *Productus semireticulatus*, *P. punctatus*, &c. (*Acad. des Sciences de Berlin*, 1861.)

My own collections have received some interesting additions from Queensland during the last year, which arrived too late to form part of the contribution to the Daintree collection.

The lower beds of these rocks, as we have seen, pass downwards to strata holding plants of acknowledged Lower Carboniferous age.

And in the upper portion of the same, though the plants just mentioned are missing, occurs a species of a genus which goes upwards into the overlying coal beds, and which because of its alliances in other countries, is held by one or two Palæontologists to carry those coal beds up to the horizon of the Oolites.

I have already written so much in denial of this determination, that, having lately obtained additional data for my opinion, I shall on this occasion content myself with enumerating the circumstances that justify this view. Did this Memoir aim at anything more than a brief and succinct statement of *observed facts*, I might again go into further argument: but it will save space to mention the facts and invite those who deny them or cavil at them, to come across the border, take off their coloured spectacles, and judge for themselves.

Those who deny the asserted age of our workable coal seams affect to rely on the assumed age of that most prominent plant—*Glossopteris Browniana*. They say Glossopteris is an Oolitic genus. "*Exactly as in the English beds the Glossopteris is associated with Tæniopteris*"? *i.e.*, in the assumed Oolitic series. To this we may reply that "*Glossopteris Browniana*" which is "*the* Glossopteris" alluded to in the above extract from the "Report of the three Commissioners on the Western Port Coal-fields," (p. 8) is a plant *utterly unknown* in Europe and America, and only known in India, South Africa, and Australia, and that Tæniopteris, which is said to be associated with it in English beds, according to Schimper, the most recent expounder of fossil botany, is a genus which has only five species, all of which are Permian *i.e.*, of Palæozoic age or of Upper Carboniferous. Even if one Tæniopteris should be found in the same beds with Glossopteris, that fact would not invalidate, but would rather strengthen my argument, since the former is Palæozoic, and the latter occurs in the coal seams below the beds which are filled with Lower Carboniferous marine fossils; it is clear that those beds and the plant they hold must certainly be Palæozoic, whatever becomes of any other part in the succession of the series or group to which they belong. It was attempted to be shown that there exists an inversion of beds at Stony Creek, where five seams of coal holding Glossopteris under 143 feet of acknowledged palæozoic marine beds occur (the fossils from which I sent down to Sir Henry Barkly, who submitted them to Prof. M'Coy), and to meet this I requested that a geologist might be sent up from Victoria to test the facts. Accordingly Mr. Daintree came, and in the *Yeoman*, Melbourne journal, No. 100, will be found his refutation of the inversion story and a full confirmation of my assertion. This circumstance is ignored by the Commissioners, as are all others that do not fall in with the imagination of certain critics in Victoria. But I may now add that Glossopteris in coal seams below the marine beds has been found in other localities, as for instance at Greta, where the coal has been reached below more than 100 feet of marine strata; Glossopteris and other plants also occurring 2 feet 6 inches above the coal. [See Sections No. 1 and No. 2 at the end of this Memoir.]

Not only so, but it is found in sandstones elsewhere, amidst the marine fossils themselves and in the very same portions of rock with the latter. So that no reasonable doubt ought to exist in the mind of an honest controversialist that "*Glossopteris*" does occur as early as the so called Lower Carboniferous strata, and therefore our coal seams have a right to be held of that age.

Now Schimper, to whom I before alluded, considers that the Indian, African, and Australian plants are merely varieties of the same *G. Browniana*. In India no marine fossils have yet been found in connection with its coal plants; and in Africa the Glossopteris is not set down to any older formation than Triassic by Mr. Tate, but even that is older (although Mesozoic) than Oolitic, to the latter of which M'Coy refers them. And if Glossopteris has a range as extensive as some other fossils which pass through three separate series of strata, why might not it pass up into Secondary rocks, without denying its existence in Australian Lower Carboniferous? *There* it clearly does not govern, but must be subordinate to the *Fauna*. But it is not alone in that position, other plants also occur therein which have as much an Oolitic facies as itself. And yet it is undoubtedly true, as is well shown by Daintree, that in Queensland Glossopteris is confined to beds that are in association with Palæozoic fauna, and that the so called Tæniopteris is found to accompany a Mezozoic fauna; and I can aver, after upwards of thirty years experience, that no marine deposits of Secondary age have yet been discovered in New South Wales; although in Queensland beds of coal occur in supposed connection with such.

There may, therefore, be two epochs of coal, as suggested by Murchison, or as stated by Mr. Carruthers, two portions of one series, without dispossessing the lower portion of its right to hold a property in a plant that may not have existed in the time of the younger part of the series. Whatever be the value or uselessness of reasoning on the point, this fact still remains—*Glossopteris Browniana* does exist in New South Wales and in Queensland in coal measures that interpolate strata full of palæozoic marine fossils, and is absent in the latter Colony where the marine accompaniments are called Mesozoic, and does not exist at all in Victoria where the Palæozoic and other marine beds are at present missing.

As to the division arbitrarily made by Professor M'Coy in a list re-arranged by him, of Mr. Keene's specimens, separating "*shale with G. Browniana and Otopteris*" from the Palæozoic beds, that excellent Palæontologist may be assured that a plant apparently the same as Otopteris? *ovata* is combined with Lepidodendroid plants near Stroud, and that at Greta, and at Mount Wingen, Glossopteris is found below his own determined Palæozoic marine fossils, the smoke from the burning seams full of the plant at the latter locality passing up through cracks in the overlying conglomerate full of Palæozoic shells, &c.

Nor does the arrangement made of Mr. Keene's collection agree with the actual facts in nature, for the Greta beds are not the uppermost with marine fossils; but beds with them lie further to the east in which Phyllotheca has occurred at Harpur's Hill and Glossopteris in the same way at Raymond Terrace.

Then, as to the "*vulgar error*" that heterocercal ganoid fishes are confined to Palæozoic beds, which any one acquainted with ordinary treatises on the subject may be supposed to understand is an error, though scarcely "*vulgar*" in the ordinary sense of that often offensively used term,—surely it may be permitted to conclude from the fact that among all the fishes discovered in our coal beds, and in the beds above the coal, not a single homocercal tail has been found, the probability is, as Sir P. Egerton has surmised, after examination of those submitted to him, that the *fishes are Palæozoic*, especially as the admission is made that "*the* "*homocercal* structure is *not* known in Palæozoic rocks." (*Report on Coal Fields.* Victoria, 1872, p. 6.)

The fact that the coal beds overlie or interpolate the marine beds in what is called "conformable order" ought to be considered a satisfactory conclusion that no break such as ought to exist under other circumstances does exist, because whether the coal measures are horizontal or inclined they merely follow the same condition in the upper or lower marine beds with which they are always associated.

The argument from the occurrence of fish remains, is met by the incidental remark that the "heterocercal ganoid fishes being of genera and species *peculiar to the locality* have no value" in determining the age of the beds in which they occur, may be met by the retort that if *peculiarity* is to be a guide in determining geological age there is an end of any certainty for such persons as affect to uphold their own theories by reference to peculiar plants; and this Professor M'Coy himself does in relation to a Scarborough plant by which he affects to guide his Oolitic determination to the exclusion of Glossopteris and its usual associates.

Respecting Palæoniscus, one of the New South Wales fishes, a passage translated from Agassiz, whose decision ought to be satisfactory, will not be out of place, considering that it meets the objection on the form of the caudal fin. He says,—"I know ten species of this genus, which appear to be limited to coal measures and the Zechstein. It might not, however, be impossible to discover traces in the *Grès bigarré*,* the Muschelkalk, and the Keuper" (*i. e.*, in the Trias); "but that which I believe I am able to affirm is, that *it does not ascend to the Jurassic formations*, of which the numerous representatives of the order of Ganoids have the tail *regular*, and never prolonged in a long point forming the upper lobe of the caudal, as takes place constantly in the genera of the earlier formations. I do not understand what were the intentions of Nature which have produced these singular differences, but it is certain that they exist, and it would be to misunderstand our duty to ignore them, or to attribute less importance to so general and constant a fact." (*Recherches sur les Poissons fossiles, tom.* 1, *p.* 43.) To this may be added, that the generality of the fishes, which are *all* heterocercal in New South Wales, are found more than 1,000 feet geologically higher than our workable coal, which those who denounce "vulgar errors" condemn to a mere Jurassic existence.

The existence of Palæozoic strata of Carboniferous age in some parts of Victoria is, as I believe, a fair assumption of the Cape Paterson Reporters, though at present they cannot prove their position by fossiliferous evidence; but the denial of that existence would hand over their whole coal territory to a formation or formations, to prove the age of which they have no more marine evidence than they have respecting a Carboniferous era. They have never yet seen a single marine fossil bed in all Victoria to justify even their adopted view of their coal belonging to the Oolitic age, which is elsewhere multitudinously fertile in marine fossils, and this, no doubt, *is* "peculiar." The Reporters on the Western Port Coal Fields notify carefully, that "it should be distinctly understood that *our opinion* respecting the age of the New South Wales coal measures is *based entirely* on the collection of rocks, fossils, and coals forwarded to us by the late Mr. Keene,

* He afterwards names P. catopterus as belonging to this sandstone. It was, however, only found in one spot, only "a few square feet" in extent, in the county of Tyrone. (*Portlock, Geology of Londonderry, Tyrone, and Fermanagh, p.* 468.)

and on the published reports on these coal fields." But even this is accompanied by a sneer at Mr. Keene's blunders in Palæontology.

On the above I would observe that, as I saw the collection referred to before it was despatched, I am prepared to say it did not completely represent the beds in the local district from which they came, and was only a partial display of the series of the strata in association with coal throughout the Colony; and that in the arrangement adopted by Professor M'Coy in the Report, most important portions of the beds are omitted. I would, therefore, attribute the "opinion" of the Board respecting the age of the New South Wales Coal," so authoritatively pronounced, to be based on imperfect data, showing that the gentlemen who have decided the question are *practically* ignorant of the true grounds of decision, clearly not having made any inspection for themselves, and totally ignoring the opinions of the host of observers who have certified to the contrary; amongst whom is Mr. Daintree, a member of the Victorian Geological Survey, the late Mr. Stutchbury, who reported thereof as well as many others who have studied the strata *in situ*, and are true witnesses against the side of the Oolitical party. In the pleadings on that side, the reliable evidence that makes against them is "burked," and a foregone conclusion is offered as if it were final—and the judgment is delivered *ex cathedrâ*, whilst numerous witnesses of the first credibility are altogether ignored. This may be prudent and ingenious, but it is *not* "*scientific*," nor is it honest, yet it helps to bring out the magnificent declaration: "We confine ourselves to the *statement* that we have *not before us a particle of evidence* indicating that the coal seams now being worked in New South Wales are of Palæozoic age." A great compliment this to persons who have laboured for years to establish truth; but they may console themselves with the reflection, that "*Préjuger est mal juger.*" Amidst this lamentable ingenuity to "tell the truth without telling the *whole* truth and nothing but the truth" and in the arraying of evidence from beyond Australia instead of collecting the whole evidence furnished from itself, there is one grateful exception which, though not entirely satisfactory, is much more so than some previous proceedings were. It would have been better to have acknowledged the change.

In the notes on Mr. Keene's specimens, Professor M'Coy, though he draws a line where it ought not to be, has changed his method of putting his old opinions about the coal itself, inasmuch as he no longer makes use of the notion which he once entertained and put in evidence before a Committee of the Melbourne Parliament. I must explain this.

On the 20th November, 1857, he was examined (as the Chairman of a Mining Commission) on the character and extent of coal in Victoria, and he asserted over and over again, that no Palaeozoic coal existed in Australia. The following answers speak to that point—

722 " (Answer). The members of the Mining Commission have an impression that, as the coal deposits to be expected there [Cape Paterson] " geologically are not the same as those of the great coal fields of England, but are of similar character with the coal deposits of New South Wales and Tasmania, therefore *it is unlikely that they will be of commercial value*; and as scientific men they would not on their own responsibility, recommend the expenditure of public money there."

727. " (Q.) Considering that the information (? formation) of the Cape Paterson Coal Fields is similar to those of New South Wales and Tasmania, you are of opinion, that as an *economic question* you would advise no further prosecution of any surveys in that locality ? (A.) That is my opinion."

744. " (Q) You would not advise the prosecution of any further inquiries for the discovery of coal ? (A.) No recommendation to that effect would emanate from myself or the Commission."

747. "Such coal fields, *i.e.* those of Palæozoic age do not exist in this country', (*i.e.* in Australia). "That is a point which I wish clearly to show, and I think it is one which has never been clearly shown to this committee before."

758. " I know you are not to expect the old Palæozoic coal fields in this part of the world."

759. " (Q.) Do you contend that the Mesozoic coal fields are not suitable for the different purposes of commerce ?" (A.) " They are not so suitable as the Palæozoic, they are not so extensive, the beds are not so thick or workable, nor is the quality so good over any workable area."

767. " (Q.) If a coal field at Cape Paterson was discovered equally good with the Sydney coal fields, would you consider it worth working ?" (A.) " My individual opinion is that it *would not be worth working*."

771. (Of Cape Paterson) " (A.) Of course the Members of the Mining Commission do not wish to attach any scientific weight to their evidence in a commercial point of view, they merely choose to say, that as *men of science*, no recommendation would emanate from them to undertake extensive works there, *because* the utmost you could expect would be such a *coal bed as you have at Sydney*." Once more ; " 769 (*By Captain Clarke.*) (Q) The Virginian coal fields of the character you describe as being similar to those here, are worked at 775 feet depth ?" " (A.) Yes ; but the beds there are *not to be compared to the palæozoic coal beds*."

No doubt the Professor was right in the last answer. But Professor Newberry is quoted, in the Report of 1872, as stating that—" Large portions of the coal basins of China, including beds both of anthracite and bituminous coal, are usually excluded from the Carboniferous formation. So large is this coal-bearing area, indeed, that when joined to the Triassic, Cretaceous and Tertiary rocks of North America, they quite overshadow the Carboniferous coals of Europe and the Mississippi Valley, and suggest the question whether the name given to the formation, which includes the most important European strata, has not been somewhat hastily chosen." (p. 8.)

Now, reconciling these quotations if we can, what is to be done with another passage in p. 9 of the Report ? In it the reporter, having arranged the order of our New South Wales beds after his own idea, says—"If their view be correct it is not likely that seams of coal, as thick and as persistent as those occurring in the Lower Mesozoic beds of New South Wales, will be found in any part of Victoria. It is to be regretted that a geological examination was not made of the northern coal-fields, during the many years the Victorian Government maintained a staff of geological surveyors, for the purpose of ascertaining by comparison the position of our beds with all the exactness practicable."

"The value of such evidence as the geologist and the palæontologist can give in such investigations as these is priceless. *They* alone can determine where the practical miner can pursue his explorations with fair chances of success."

Thus speaks out the modern Delphi—but what becomes, after all, of the expectation of the anticipated Mesozoic coal beds of Victoria, and what must Mr. Daintree, who was one of the staff spoken of, think of the way in which his success in carrying out the investigation recommended at Stony Creek is rewarded when that very important work is totally ignored by the Palæontologist of the survey, by whom all the specimens collected, sent to him by me, were examined, and who now has had his eyes so far opened as to acknowledge that some "Palæozoic" coal does exist in New South Wales?*

As to the fact of changing an opinion on conviction being wrong, he who so changes is not to be taunted with it unkindly, and I do not advance it except to acknowledge that so far as the

* In reference to the above remark the following passages from "Geological Notes, with Plan and Section, by Richard Daintree, Field Geologist, Victoria," may be properly cited.

"From Newcastle to Stony Creek is but a short trip, and as these are the sections on which Mr. Clarke bases his evidence of the Palæozoic age of part, at least, of the New South Wales coal seams, it is one of the necessary pilgrimages for the wandering geologist in search of truth. What I saw there I will state in as few words as possible. I saw three shafts on Mr. Russell's estate—ladder shaft, working shaft, and 200 feet shaft."

He then gives his measurements, which are not material to cite in this place, and goes on:—

"When the details of these shafts were first made known by Mr. Clarke, as a proof of the Palæozoic age of the coal, Spirifers, Fenestella, &c., being found in abundance, and Glossopteris associated with and below the coal, it was suggested by Professor M'Coy that the data given by Mr. Clarke showed the existence of a fault between 'working' and '200 feet shaft,' and that possibly to this fault the reversion of beds might be due, but the Palæozoic character of the Fauna was not called in question.

"This error arose from taking the absolute distance between the shafts (360 feet), instead of the reduced distance to the line of dip of 280 feet.

"Referring to the extension of Russell's coal seams to the Northern Railway, unfortunately at a point where no marked bed of Russell's series can be absolutely identified," [but at that point may be identified both plants and marine fossils and traces of coal in the strata there disturbed] "we have an apparently unbroken series of strata dipping in the same direction, and at about the same angle, as those in Russell's coal pits, extending from a point at 19 miles 75 chains from Honeysuckle Flat to 21 miles 37 chains from the same place, the beds furthest to the eastward dipping at a greater angle.

"This affords a thickness (taking the angle of dip at 16 deg.) of 2,365 feet of strata, abounding in fossil fauna from bottom to top, very low down in which coal seams with Glossopteris occur.

"Fossils from each of the cuttings on the Railway and from Russell's shafts were procured, that Palæontologists may satisfy themselves of their European parallel.

"If it be admitted that the Fauna found in the upper strata of these shafts is Palæozoic, then these coal seams at least are Palæozoic, and Glossopteris has a much lower range than has hitherto been assigned to it, except by Mr. Clarke.

"Neither does there seem any reason why Mr. Clarke should not place the Newcastle coal seams (his No. 3 Carboniferous group) in the upper portion of this Stony Creek group,

Professor has gone, he deserves respect and honour for the change. My only complaint is that he has *not gone far enough*; though after what he and his colleagues announced in the examination above referred to, respecting the sole Mesozoic character of our New South Wales coal, it is refreshing to find him writing in these terms of the Greta and Anvil Creek coal seams. "The beds from "*k.*" to "*n.*" (referring to his re-arrangement of Mr. Keene's specimens) are clearly the marine Palæozoic Carboniferous rocks, and *the coal found with them resembles the coal of the southern coal-fields of Ireland of the same age.*" But he adds without compunction or authority:—"Neither this collection, nor the sections, nor Mr. Keene's collection in the Melbourne Exhibition, bear out the notion that the Glossopteris and Phyllotheca alternate with the marine Palæozoic shell beds." Now had a visit been paid by him to the localities of Rix's Creek and the rest, or to Anvil or to Stony Creek, or to Mount Wingen, such an assertion would not have required fresh denial from me; and to jump from the Wallsend seam to Rix's Creek, and Anvil Creek, without any examination of the section of the intermediate localities, or to deny the existence of Glossopteris at those and other places among the marine beds which are so interpolated, is to do away with the whole merit of such a section as the "notes" pretend to represent.

---

no known unconformity existing, since no Fauna or Flora typical of the Mesozoic period has, I believe, yet been been found in the said No. 3.

"This brings me to the consideration of Mr. Clarke's present arrangement of the Carboniferous series of New South Wales.

"*First.*—'Wianamatta' beds, with insignificant coal seams, the upper beds of which are the probable equivalents of our Otway, Bellerine, and Wannon beds, in which Glossopteris has not yet been found.

"*Second.*—'Hawkesbury' beds, with insignificant coal seams; no Glossopteris. To this series Mr. Clarke refers the Grampian sandstones of Victoria, though Mr. Selwyn places them with No. 4. (By Grampian sandstones I mean the beds constituting the Sierra.)

"*Third.*—'Carboniferous beds,' containing the workable coal seams, with Glossopteris, by far the most abundant fossil. In the lower portion of this series four (? five) known coal seams are interpolated with strata containing a Fauna similar in character to that found in the Carboniferous limestone of Europe.

"*Fourth.*—'Lepidodendron beds,' not associated with coal seams, as far as yet known.

"If this arrangement is correct—and my experience as a field geologist is entirely in its favour—it is of great practical value to us in Victoria in the search of workable coal seams, &c., &c., * * in the hope of finding the Glossopteris beds. It points unfavourably towards the Tæniopteris and Zamites-bearing beds, which we have hitherto regarded as our coal-producers, but which as yet have yielded nothing better than the Cape Paterson cams.

"Four thousand feet also of these same beds have been tested by boring in the Bellerine District, and have yielded nothing approaching a workable seam.

* * * * * * * * * *

"All the facts that we have to guide the field geologist in Victoria, in his search for Clarke's No. 3 carboniferous beds (containing the workable seams of New South Wales) are these—that they are very low down in the Carboniferous series; that the lowest beds contain a Fauna nearly allied to the Lower Carboniferous of Europe; that Glossopteris is associated with all the coal seams, and is the most common and characteristic fossil of the said No. 3. This peculiar Fauna or Flora has not yet been observed in Victoria."

(From *Yeoman and Australian Acclimatiser*, August 29, 1863, No. 100, *published at Melbourne.*)

It will be unnecessary to point out to any unprejudiced reader how Mr. Daintree's "Notes" cited above, known as they must have been to the "Reporters on Coal Fields, Western Port," nearly nine years before, contrast with their lamentation in the year 1872, about the "non-comparison" by *Victorian surveyors* of the position of the coal beds in the two Colonies, "with all the exactness possible."

I will quote here an additional testimony to the facts already declared, respecting the interpolation of our Glossopteris coal in the marine beds. Mr. Odernheimer in his final report to the Australian Agricultural Company, says:—"The lowest coal seam at Wollongong, rests on older Spirifer sandstone, and is covered by sandstone, with Pachydomus shells and a few spirifer." (p. 88.)

I have paid more attention perhaps, to the Report on the Western Port Coal Fields of 1872, than it deserves; but as it contains specific allusions to myself, and in fact is an attack on the evidence I have conscientiously given on the subject of New South Wales Geology, it is only just to that Colony to show that the conclusions arrived at in that Report are "based" as much on personal ignorance respecting our territory, and a pre-determination to disbelieve the statements of men quite as much entitled to be believed as the reporters in Victoria themselves, as on anything else. I am thoroughly persuaded that if such personal investigation on his part had taken place, an old correspondent and assumed friend of my own would not have dealt with my writings as he has done.

The advocates for the Oolitic (or as now called Mesozoic) age of our coal plead the cases of Richmond in America, and India as well as China; Africa is unnoticed. It will be fitting to produce evidence on each head.

As to China, Mr. Pumpelly is the only authority quoted by the Victorian Board, who make him to have in 1862-65 found in the coal beds *fossils* proving that "those beds are *geologically* of the *same age* as the *Victorian, New South Wales, Tasmanian, and New Zealand beds*," p. 8, and Professor Newberry is quoted as identifying "these fossils as those characteristic of Triassic *or* Jurassic ages." In the *Ocean Highways* for Nov., 1873, Baron von Richthofen says, the Pumpelly observations were only very limited in extent, and his map an hypothetical one made up from native reports, "in which he attempted to exhibit among other data, the distribution of the coal measures in China." "The favourable result at which Mr. Pumpelly arrived, in respect to the great extent occupied by coal-bearing strata in China was modified in some measure by the *somewhat unsatisfactory conclusion* drawn by him, from the determinations by Dr. Newberry of a *few vegetable remains*, that all Chinese measures are of the same age as the *Triassic* formation of Europe" (p. 311). What is there herein of "*Jurassic*" or "Oolitic" coal? The coal of China, however, found a far better qualified expositor in Baron Von Richthofen himself, who from 1868 to 1872, made journeys nearly all over China, and found coal-fields of enormous extent in many districts, nearly every one of which he personally visited, as he tells us in various publications.

He mentions one seam of Silurian age; several others in Devonian strata; but he adds "*the great bulk of the most widely distributed and most valuable coal-beds are proved by numerous and very characteristic marine fossils to belong to the true carboniferous.* After the close of that epoch the deposition continued without interruption through the Permian, till probably towards the *close of the Triassic epoch.*"

These are his own words, and he justifies his determination of epochs by informing us, that "he first determined with some accuracy the geological age of the sedimentary formations by a great number of prolific fossiliferous localities." Nowhere in this account of his do we find mention of Oolitic or Jurassic coal. So that really China should not be quoted to uphold the "*same group as the Cape Paterson series,*" (Report p. 5). Rather might it uphold the coal of New South Wales. If marine fossils are "necessary," none exist in Victoria as we have already seen and as the Report allows. "The coal measures of Richmond, Virginia"—the Report also says—"are stated by Sir C. Lyell to belong to the lower part of the Jurassic group," (p. 8). Well, he did *once say so*, but he found he was wrong, and so he placed them finally in the *Trias*, Professor Heer considering that the plants "have the nearest affinity to the European Keuper." (*Student's Elements of Geology*, 1871, *p.* 362.)

In Africa, the association of the genera Glossopteris, Phyllotheca, and Dictyopteris, "affords some evidence of Mesozoic affinities" says Mr. Tate, who, nevertheless, shows that the shales in which they occur are not Jurassic, but Triassic. (Q. J. G. S., XXIII, p. 142.) Palaeoniscus and some of the reptiles and an encrinital stem, might refer these Karoo beds to a lower position still. Mr. Tate admits the analogy is with the Keuper (p. 169).

On a former occasion, I entered upon an inquiry as to how far the coal fields of India were parallel with those of New South Wales, and how far they corresponded with the view of a Palæozoic age for the latter, as shown by the determinations of Dr. Oldham, the able Superintendent of the Geological Survey of that country. On this occasion I may mention that, being desirous of ascertaining whether any change had taken place in the views of that excellent geologist on the question of age, I wrote to him to request he would kindly satisfy my inquiry. On 2nd June, 1874, I received his reply, dated 2nd April of that year, so that it may be taken to give the actual present state of the Indian Coal Fields history. I shall, I believe, involve no breach of confidence in quoting his own words, which will save the necessity of again searching the Memoirs and Records of the Survey:—

"We have seen," he says, "no reason whatever to alter our views with reference to the age of our Indian coal rocks. The plant evidence is tolerably conclusive with us. Our *upper* beds, which contain thin patches and threads of coal (and which we call RAJMAHAL formation), we have established, by a careful research in Cutch, to be *Upper Oolite*. These are characterized by an abundance of Cycadea and Tæniopteris, but not a single Glossopteris has been found. Then we have the group we call the PANCHET System, with no Cycads. Schizoneura (a plant first described from the Vosges), &c., and with them Labyrinthodont and Dicynodont reptiles. No Glossopteris here either.

"Then below these, with slight unconformity over the coal rocks, in which, observe, we find *Glossopteris Browniana* abundant; and this holds through the several thousand feet of thickness, occurring in all.

"At the base we have a small thickness (relatively) of the TALCHEER System, in which Cyclopteris shows, but no Glossopteris.

"Unfortunately we have as yet no animal remains from our coal-rocks. Notwithstanding this, in connection with your evidence from Australia, and bearing in mind the perfectly established identity of the Glossopteris, even in its varieties, and the *equally established fact that Glossopteris has never been found in Europe*, and therefore gives no clue or index to age from European determination, I cannot come to any other conclusion than I have done, that our coal in India represents *the latest portion of the Carboniferous of Europe, and the gap between this and the Permian*; or, I would say, in a broader sense, *the latest part of the Palæozoic time.*

"I read Daintree's paper with much interest, and think he has done much to clear up some of the difficulties.

"But so long as some fancied analogies with regard to fossils are allowed to sway the mind, there can be no agreement of opinion."

"The Glossopteris of Australia and India are identical. We have every variety, as described from your beds, and no one could hesitate to admit that the beds are similar also. All these Glossopteris beds must be admitted to be of similar relative age in both countries. It proves *nothing* as to the age *relating* to *European systems*. You know better than I do the amount of co-existing evidence as to age which you have established in Australia.

"In India it is this, in a few words:—

(3.) *Above*—A system of rocks, with abundance of Cycads, Tæniopteris, Pecopterids, &c., &c., truly Oolitic with their threads of coal.

(2.) *Next*, separated by considerable time beds with Schizoneura, Pecopteris (*no* Tæniopteris, *no* Glossopteris), Labyrinthodont, and Dicynodont reptiles, the analogies of which are Permian or certainly Lower Triassic (*no* coal).

(1.) *Next*—The coal rocks also separated by unconformity, though slight, which have abundance of Glossopteris and also of Schizoneura of different species—as yet no animal remains.

"There are thus three distinct floræ with no species common to each You can draw your own conclusions.—T.O."

In the above remarks of my distinguished friend are some hints that will not fail to be of use in relation to New South Wales, as well as to other parts of Australia, and it is satisfactory to myself to have so much confirmation of my own views. Though it is true that Glossopteris, not being a European plant, does not confer any claim on itself to designate the age of our coal beds, yet assuredly as it occurs in the Lower Carboniferous beds as well as in the Upper coal measures, it does bear on their association with the greatest force, and the two series of beds must be nearly of the same relative age. That age as pointed out by Dr. Oldham, and as I have all along stated, must be Palæozoic.

As to the coal beds with no Glossopteris they will go with rocks of a more recent date, and there can be no objection to class them in the age of the Secondary fossils with which they are associated. Professor M'Coy himself admits:—"That on mere fragments of leaves or other most imperfect or ambiguous material no generic nor even ordinal characteristics should be founded." (*Observations on Vegetable Fossils of Auriferous Drifts, by Baron von Mueller*, 1874." p. 11.) But this argument does not apply where fragments even of the same plant occur in two series of beds. Resting on, or passing into each other without a break, they would assuredly show that such beds are intimately related.

If the idea be abandoned (and *there is no real authority for it*) that Glossopteris is an Oolitic plant, and if it be admitted that a Fauna has more weight than a Flora, and that it is most probable that a floral identity never existed during the same epoch at the Antipodes of the European Oolitic area, more reasonable will appear the position assigned by me to the New South Wales workable coal-beds.

Is it more remarkable that *plants* held to be of Mesozoic age in Europe should be found at the Antipodes in a Palæozoic formation, than that usually considered Mesozoic *mollusca* should be found in a similar formation? And the latter is not merely a conjecture but a fact, attested by Palæontologists of eminence. For instance, Münster in 1841 found the three genera *Ammonites*, *Ceratites*, *and Goniatites* in one and the same bed belonging to the St. Cassian rocks of Austria; and now we have Dr. Waagan, of the Geological Survey of India, proving to us that the *same three genera* have been found in the same bed together on the Salt Range, in the society of Productæ, Athyris, and other well-known Carboniferous fossils, pointing out that the Ammonites is there a Palæozoic genus, which he places either in the upper part of the Carboniferous, or as Dr. Oldham considers our disputed coal beds may be, about the limits of the Permian and Carboniferous formations.

Whilst discoveries such as this are being made from time to time, what obstinate perseverance is it, to continue to maintain that even the stereotyped determinations of palæontologists are incapable of amendment. (For Dr. Waagan's description and figures, see "*Memoirs of the Geological Survey of India*," vol. ix, part 2, p. 351. See also Lyell's "Elements," 1865, p. 436, and "Student's Edition," 1871, p. 358.)

Nowhere in N.S.W. has there yet been found in association with the plant beds any marine Fauna but one, which M'Coy and all other Palæontologists admit to be *Palæozoic*. Schimper, in his recent powerful work (*Paléontologie végétale*), does assume on the statement of reporters that Glossopteris occurs in the Oolitic formation of the Rajmahal Hills of India, but Dr. Oldham, the skilful Director of the Indian Survey, declares that its officers have not "*been able to trace, among several thousand specimens, a single representative of the genus Glossopteris* from any part of *these upper or* RAJMAHAL *beds.*" (See also his statements above.)

If then, that series of beds be considered Mesozoic on other evidence, and if, as is shown, Glossopteris belongs to a lower group or formation there is here an enormous thickness of fossiliferous strata, in which the fossils (as before stated) gradually pass down to the Devonian. The opposition to this determination arose from a preconceived idea that strata bearing Glossopteris could not be Palæozoic, and therefore, that the upper coal measures of Newcastle had no right to be considered older than *Oolitic.* But whilst these upper measures produced a fish of undoubted Palæozoic character (Urosthenes *australis*); Cleithrolepis *granulatus*, Myriolepis *Clarkei*, and other Icthyolites, examined and determined by Sir P. de M. G. Egerton, Bart. to be Palæozoic, were found by me in 1865 1,000 feet higher, and of these, photographs were exhibited at the Paris Exhibition of 1866–7, and previously at Melbourne the specimens themselves; on which occasion Professor M'Coy reported that their general aspect was that of Triassic or Permian fish to which latter (Upper Palæozoic), Sir P. Egerton refers them. Alethopteris *lonchitica* and Adiantites *eximius*, both of which occur in our New South Wales beds, may be held to have as great weight as Glossopteris, seeing they occur in an unbroken series of beds holding true Lower Carboniferous marine fossils, and are, I believe, considered to be of Carboniferous age.

After the evidence from Queensland, and the admission that the plant does not exist at all in Victoria (where all marine strata are missing also), *Glossopteris* cannot be cited from that Colony to assist in proving New South Wales Newcastle coal to be Oolitic; and there are sections on the Bowen River (full 1,000 miles from Sydney), in which the whole history of the coal-beds may be read off without error.

A conclusive opinion has been offered on this question by Dr. Julius Haast respecting the occurence of marine and plant beds of the same age as ours in the Malvern Hill District, Canterbury, New Zealand, who says, in October, 1871 (N.Z. Geological Survey Reports on Geological Explorations, during 1871 2), that on the west side of Mount Potts, Upper Rangitata, there are "different species of Spirifera; besides them there are species of Productus, Murchisonia, Euomphalus, Nucula, Orthis, and Orthoceras. Most of these shells, of which some broad winged Spirifers are very numerous, are according to Professor M'Coy, of Melbourne, identical with Australian fossils, and are of Lower Carboniferous or Upper Devonian age." "Other beds," he adds, "of equal importance occur in the Clent Hills, in which I gathered a rich harvest of fossil ferns, mostly Pecopteris, Tæniopteris, and Camptopteris" (this, however, is not found in New South Wales) "which, according to Professor M'Coy, are of Jurassic age identical with beds belonging to the New South Wales Coal Fields, and although I believe this Clent Hill series to be somewhat younger than the Spirifera beds, I demurred to this definition, owing to the fact that the position of the strata and the character of the rocks of which they are composed have quite a Palæozoic facies."

"Since then it has been shown, and as I think with conclusive evidence, that both fossiliferous strata, the Spirifera and Pecopteris beds occurring together in the New South Wales Coal-fields, are of the same age, and alternate with each other. The occurrence of Tæniopteris, which hitherto has been considered only of Secondary age,* seems to speak against a Palæozoic origin; however, I may point out that, the same objection was made to the Glossopteris in Australia, but which has by overwhelming evidence been shown to be also of Palæozoic age. I do not think that the fragment of a leaf, however distinct, can unsettle all that stratigraphical geology has proved to be correct." (p. 6-7.)

Some recent researches made by me, with a view to the consideration of this question of age, render it far from improbable that a series of beds has been swept off the coal measures by denudation, in which marine beds may have overlain the now existing strata, just as in a lower horizon they do still at Stony Creek, Anvil Creek, Mount Wingen, and in other localities. The facts that the present coal seams range in elevation along the coast, from below the sea, to between 200 and 300 feet only above it, and that to the westward they reach an elevation of upwards of 3,000 feet, still preserving the same plants as below, and with an equal almost horizontal level (except in cases where local derangement has occurred from special elevating forces), and moreover, that similar seams occur at various other elevations between those mentioned, induce me to consider it possible that there has been a sinking along the coast line, allowing denudation to operate.

At present this hint may not be worth much, but hereafter more may come out of it. I ought also to add that between the Hawkesbury rocks and the coal there is often a series of beds belonging to the coal measures in which marine Palæozoic fossils are stated to have been found.

In the sections published some years ago by Mr. J. Mackenzie and myself, and in subsequent sections by the former, as given in his Report to Government, it will be seen that the number and thickness of the seams vary considerably in different localities. The former circumstance may be accounted for by the fact that the

* Schimper says (tom. 1, p. 600) of the genus Tæniopteris—"*Ces Fougères paraissent être propres au terrain houiller supérieur et au permien*," i.e., they are Palæozoic.

beds in the coal measures were deposited over various older formations, some here, some there, which occur at different levels, so that some of the strata are missing in a few of the localities, and all are seldom seen together. Thus the coal series at the height of 3,000 feet does not contain so many seams as near the sea level. And, perhaps, in describing them it would be preferable to separate the deposits into various local basins or saucers; though the conditions of a true basin can only be exhibited on the large scale.

It is at least certain, that in the Western districts, though many of the conditions of the Newcastle and Illawarra beds exist, there are found certain fossils which are not found in the latter, and which would lead to the presumption that, as we ascend in height above the sea we find the introduction of genera gradually approximating to a more recent epoch. For example, the upper beds of the Lithgow Valley coal measures contain a fossil which I first collected in 1863, and of which Mr. Wilkinson has lately gathered some striking examples. These coniferous fossils consist of stems and branches ending in Strobilites. Professor Dana, to whom I sent specimens, informed me that he had never seen such before. To me they appear not unlike the Strobilites from the *Grés bigarré* of Soulz-les-Bains, in the Vosges, figured by Schimper and Mougeot (*Monographie des Plantes fossiles de la Chaine des Vosges. Leipzig*, 1844, *tab.* xvi, p. 31.)

In another direction, viz., on the Clarence River, there is a patch of coal measures in which there are forms resembling that of Walchia, with abundance of fragments of a plant common in the Mont d'Or coal measures of New Caledonia, together with plants that have a Tæniopteroid character, but are not Tæniopteris. On the other hand, on Bundanoon Creek, in the county of Camden, there is a Dictyopteris.

As far as some of the plants are concerned, it may be admitted that they are in an unsatisfactory condition at present; but the balance in favour of a "Carboniferous" age for the Glossopteris beds is, to my mind, conclusive.

With Dr. Oldham's arrangement in view, as given above (p. 174), there is no difficulty in admitting that in New South Wales there might be as many groups as in India, each younger than the other, without underrating the antiquity of the oldest.

With respect to the uppermost Palæozoic rocks, Mr. Etheridge states that "The occurrence of Permian strata has not been confirmed in Australia," which Professor M'Coy surmised from Productus *calva* and *Aulosteges* or *Strophalosia*, submitted by me to the latter naturalist in 1860. It is but just to Professor M'Coy to explain, that they were collected in 1856 by Mr. Gregory on the Mantuan Downs, and forwarded to me by him in 1860.

So far, then, the question about the age of some of the Australian coal must be considered as settled; and if, as in Illawarra, the coal beds overlie the marine beds, as they do also in the Fingal district of Tasmania, it would appear that all these separate occurrences belong to one thick series, in which marine beds and fresh-water beds interpolate each other. But, assuredly, in that case, the arrangement adopted must express the order as follows:—

1. Upper coal measures.
2. Upper marine beds.
3. Lower coal measures.
4. Lower marine beds.

So far as I know, the latter rest frequently on a conglomerate, which in Tasmania I found to contain undoubted Carboniferous fossils.

Since the Exhibition of 1862, on which occasion, in a paper on the Coal Fields, I noticed the occurrence of oil-bearing cannel coal at the foot of Mount York, and at Colley Creek in the Liverpool Ranges (not on eastern waters), the former has been in great request for the purpose of producing illuminating oils; and the produce has been brought into the market. In the former locality, and in Burragorang, I have made some researches which have satisfied me that these can only belong to the upper coal measures. At Burragorang the blocks of cannel are found in an intermediate position, between the top of the coal measures and the upper marine beds, which (if the overlying measures themselves do not) certainly bear the very strongest resemblance to a part of the Hunter River series.

In Illawarra, also, there are shales which are above that geological position, and which produce oil for illumination, but are not of the peculiar character of the cannel at Mount York, which in a great degree, resembles the Bog Head mineral of Scotland, only it is more valuable. The character of this substance is such as to justify its being considered a species of Bathvillite or Torbanite, in consequence of its colour and woody condition.

It has unquestionably resulted from the local deposition of some resinous wood, and passes generally into ordinary coal, many portions of the same bed exhibiting the unmistakable features of the latter and the impress of fronds of Glossopteris as plainly as they are shown on ordinary coal shale. This hydrocarbon varies somewhat in composition; and (as at Colley Creek) is frequently filled with quartzose particles, showing that it was deposited in a shallow pool, to which sand was drifted perhaps by the wind.

At Reedy Creek, now called Petrolia, there is a band of thin and very elastic substance of this kind, separated from the thicker bed below by a parting of white clay.

Varieties of this mineral occur in the Grose River, at Burragorang, on the Colo, on Mount Victoria, and in one spot in Tasmania behind Table Cape, on the southern shore of Bass's Strait, as well as in other localities in other Colonies. Presuming that the origin above suggested is correct, viz., the occasional occurrence in the ancient deposits of trees of a peculiar resinous constitution, there is no anomaly in finding in one spot a mere patch amidst a coal seam (as is the case at Anvil Creek, on the Hunter River), or thick-bedded masses of greater area as in the coal seams of Mount York, or of American Creek in the Illawarra, depending on the original amount of drift timber.

In the section presented by the escarpment on the left bank of Cox's River, below Pulpit Hill, at Megalong, there are two beds in which this hydrocarbon exists.

Some time since specimens of this, together with others from the Illawarra, were taken to America by Mr. Consul Hall, and were subjected to examination by Professor Silliman. The result was afterwards published in the *American Journal of Science and Art*, under the name of Wollongongite, an accidental misnomer (as I have elsewhere pointed out), inasmuch as I have Mr. Hall's written assurance that the specimens examined by Professor Silliman did not come from the Illawarra, but from the western sections at Megalong and Reedy Creek.

Professor Silliman shows that this material, as tested by him, has an illuminating power very much greater than any other yet known. It would be invaluable if it existed in sufficient quantity to meet all demands upon it. As it is, there are two separate oil-producing works (one on American Creek, the other in Petrolia), which are now employed in making mineral oils of reasonably good quality, though both inferior to the product described by Professor Silliman.

It has been an object of inquiry whether Petroleum springs exist in New South Wales. Such have been reported from the Corong in South Australia, and from Taranaki in New Zealand, and from Victoria. The former is, we learn, a mistake, being probably at a point where certain animal substances have decomposed. In New South Wales there are also two localities, known to me for many years, in which *a nitrous product* exudes; and there are two or three in Western Australia of the same kind, which I examined. Nothing of value has as yet been found.

Supposing the truth of the conjecture respecting the formation of Torbanite and its allies from chemical decomposition and changes of resinous kinds of drift timber in the masses now transformed to coal, the occurrence of such a mineral is not necessarily confined to coal-beds of one epoch; and thus we find Dr. Hector reporting on the occurrence of a hydrocarbon in New Zealand, from what he deems a Secondary formation, intermediate in volatile matter between those of Torbane Hill and New South Wales, the latter having by far the greatest amount, with much less ash than the former.

Mesozoic or Secondary Formations.

It has been supposed that I have a dislike to rocks of Mesozoic age; but the endeavours made by me to bring to light their existence in Australia, (see Mr. Daintree's notes, and Mr. C. Moore's paper in Q.J.G.S., vol xxvi, 226–261) ought to save me from any imputation of that kind. I can only say, that whether I have been mistaken or not in any given case connected with the geological epochs of Australasia, it is not from want of honest devotion to the cause of truth, nor from a desire to hold my own without reason against those who differ from me, that I have in so many publications during more than thirty years of earnest inquiry, defended what I conscientiously believe.

With this admission I may go on to explain, that though I hold our worked coal seams, which now extend lower than the Newcastle strata, to be Palæozoic, there are in Queensland, Victoria, and New South Wales, deposits of coal from which the characteristic plant and its associates appear to be excluded.

The rule, I think, in such a case as that before us, should be laid down, that plant remains *by themselves* prove very little as to the uncompared age of any formation, but when *associated with marine fossils, whose age is determinable*, they must go with that formation of whatever age it may be; for although plants may be swept into the ocean at any period of their existence, they could not be bedded in the same masses of stone formed in the ocean and amidst the marine fossils, without belonging to the epoch of the latter.

Such is the case in Australia with Glossopteris, and perhaps some others; hence I claim for that at least a Palæozoic age. And so with those described by Mr. Etheridge and Mr. Moore (in the Memoirs above cited) the Mesozoic *marine* fossils prove the plants to be of that epoch; and when the same plants occur in strata which can be referred to a Secondary formation, and in such also as are carboniferous, it may be readily granted that they are common to the two. But in the case of Glossopteris no indication is at present producible of its existence in the later formations.

We may therefore refer certain deposits in Queensland, in parts of New South Wales, or the coal series of Victoria, to Mesozoic (not Oolitic) times, without trenching on the Carboniferous indications. I do not profess to know—and I know no one who is able to tell me—why such arrangements exist (especially as Mr. Carruthers' doctrine is true, that Tæniopteris and Glossopteris are akin in structure) as place plants very much alike in some respects in different epochs, without confusion, when also the position of the strata is what is called "conformable."

It is no logical argument to say that, because there may be *great* deposits of coal in China or America or Great Britain, that are not what are called *Carboniferous*, therefore, there ought to be such in Victoria, when we all know they do not exist there, or that the same citations would bear out the assertion, that the New South Wales workable seams are also Secondary; nor can the adroit alteration of the expression *Oolitic* into *Mesozoic*, prevent our considering that the general term was adopted for the more specific one, because those who used it so were aware that they had made some kind of mistake, and did not like to own it.

Now, there are no *known* Oolitic marine fossils in all New South Wales; and the Oolitic or Jurassic fossils are of such extent and variety in all countries, wherever the regions in which they occur have been explored, that to put the identity of such formations on a few *plants*, that may after all have no strict claim to decide in the cause, would appear to me a very questionable proceeding.

If, for instance, the fishes found by me in the Gib Tunnel Range, near Nattai, are of a "Triassic or Permian" facies, according to M'Coy, and are Permian according to Egerton and Dana, why should the beds in which they occur be set down as Oolitic or Jurassic, instead of "Triassic or Permian"? Sir P. Egerton has shown that, with Palæonsicus, occur other genera, closely related to Pygopterus, Acrolepis, and Platysomus, all either Upper Carboniferous or Permian genera in other parts of the world.

Then again, why should the Urosthenes of Dana, from a prominent part of the Newcastle local beds be left out of the same category?

Is not the view that all these beds, ranging in succession, one over the other, and being all, as I believe, of fresh water origin (for the Hawkesbury rocks contain plants, but no animal remains except fishes), have a common relationship, and yet with no pretext for a Jurassic origin on the score of animal co-existences of that era? When we consider that the fishes alluded to occur at different altitudes, and are all heterocereal Ganoids, we must conclude that there have been physical disruptions, and that there are gaps in the succession occasioned by following denudation, or that there have been repetitions of strata now no longer traceable. For instance, the fish beds are at Cockatoo Island, 16 feet below the sea; at Sydney, less than 100 feet above it; 100 feet at Paramatta; 250 feet above it at Campbelltown; 780 feet above at Redbank, near Picton; 1,100 feet on Razorback; 2,300 feet at the Gib-Tunnel;

and 3,450 feet on the Blue Mountains; the lowest two stations and the highest being in the Hawkesbury, and the others in the Wianamatta beds above the Hawkesbury; whilst at Newcastle, the Urosthenes was the deepest below the sea.

As necessary to explain still further the succession of strata, I introduce here some additional remarks on the Supra-carboniferous rocks in the province of New South Wales.

Over the uppermost workable coal measures of that Colony, is deposited a series of beds of sandstone, shale, and conglomerate, oftentimes concretionary in structure and very thick-bedded, varying in composition, with occasional false-bedding, deeply excavated, and so forming deep ravines with lofty escarpments, to the upper part of which series I have given the name of Hawkesbury rocks, owing to their great development along the course of the river-basin of that name. These beds are not less in the coast region than from 800 to 1,000 feet in thickness, containing occasional patches of shale, with fragments of fronds and stems of ferns, a few pebbles of porphyry, granite, or slates, and assume in surface outline the appearance of granite, from the materials of which and associated old deposits they must in part have been derived. On the summit of the Blue Mountains, as along the Grose River, the thickness of the series is very much greater. Patches of very small area contain coal, carbonate of iron, and other representations of miniature coal measures.

Towards the base, patches of purple shales are frequent, and many ferriferous veins, with specular iron, hæmatite, ilmenite, graphite and other minerals, sometimes occur.

In places, as about the "Yellow rock," near the Upper Wollombi River, in Ben Bullen and above the deep excavation of the Capertee amphitheatre, salt and alum are found in cavities formed by decomposition, and in other places, as at Bundanoon Creek in the Shoalhaven District, at Appin, and on the Bullai escarpment of the Illawarra, and at Pittwater, north of Sydney, stalactites have been formed under similar circumstances.

There is an enormous mass of brown iron ore highly carbonised, partly worked at Fitzroy, near Nattai, another on Brisbane Water, and a smaller, on the coast, a few miles north of Sydney, and other similar patches in intermediate localities. These are in part associated with specular iron, which occasionally lines the joints of the sandstones close at hand with well-formed crystals.

The uppermost beds of this formation, especially where they become conglomerates, exhibit isolated summits imitating ruined castles, and have thus been traced by me at intervals all along the escarpments to the westward of Sydney, from the latitude of the Clyde River to that of the Talbragar, and in certain localities within the longitudes of that line and the coast. In the deep ravines of the Grose and Dargan's Creek, the one eastward and the other westward of the Darling Causeway traversed by the Western Railway Line, the slopes are studded by fantastic pillars sculptured by denudation and decay into imitative architectural forms. Similar forms cap the extension of the coast range to the head of the Goulburn River.

This group of Hawkesbury rocks, very *improperly denominated* by some writers "*Sydney sandstone*," (which is not a type of the whole formation, and is borrowed from the first explorers, who had never gone far into the country, besides involving a confusion with the sandstones of the Sydney Coal-field of Cape Breton in North America), is surmounted by another group, or series of strata, called by me Wianamatta beds, which are, if not in all places, generally conformable with underlying, pot-holed Hawkesbury rocks (as is well seen at Myrtle Creek, near Picton), but are connected with the underlying group by means of shales holding ironstone nodules, abundant fossil wood, fish remains and freshwater shells allied to Unio, Cyclas, &c. These beds pass upwards into highly calcareous sandstones, which also contain plants, stems, and leaves, and cone in cone carbonate of iron. These harder beds also contain Entomostraca, some of which were long ago submitted by me to Professor Rupert Jones. The fishes were examined by Sir Philip Egerton, who considers them to be Permian, as before stated. The last specimen of fish from the Palæoniscus beds, reported by me to Sir Philip Egerton, was a portion of a jaw of a fish whose teeth were of a Saurichthyian type, but the learned Icthyologist considered it also to be Permian.

Could I have procured the remains of fishes that have been reported to me from beds below the upper coal, and of the finding of which there is pretty good evidence, we might have been able to show that the same genera that we find ranging from the Wianamatta down to the coal measures of Newcastle, all through the Hawkesbury series, occur still lower.

A Palæoniscus, found since my discovery in 1860 was exhibited by the Surveyor General (who gleaned after my harvest), in the Exhibiton of 1875 at Sydney, and a specimen of Cleithrolepis found in a railway cutting on the Blue Mountains was shown by Mr. T. Brown, M.P., to whom it had been given by the finder after I had had it photographed. These formed part of the collection exhibited by the Mining Department.

Whatever may be the age of the Hawkesbury and Wianamatta beds, they contain only patches and threads, but no seams of coal. In the former the coal occurs in the sandstone in little threads a few inches or perhaps feet long, and an inch or two in thickness, and such may be seen in the walls of buildings in Sydney.

From the same beds of sandstone also I possess specimens containing ferns, like Odontopteris; and from the Wianamatta beds columnar and pisolitic iron ore, with many fragments of stems or leaves of ferns, different in species from those of the coal measures; but in neither series is there any Glossopteris or any coal *seam*. The sandstones of the Wianamatta beds are finer in grain than those of the Hawkesbury, but very much more compact and heavier, and often calcareous. The tints of the latter are *poikilitic*, darkening from exposure, and exhibiting imitations of landscapes sometimes of striking character. The semi-crystalline fragments of quartz, and the disposal of colours (suggesting the idea of the action of gases removing the ferruginous tint in places) have caused me to believe that some transmuting agency has affected large areas of the Hawkesbury rocks. The glistening of the crystalline quartz particles reminds one of the same character observable in the millstone grit of England. It is impossible to understand how considerable masses of the sandstones could have received such a present structure without the metamorphism suggested; for the crystalline facets are quite unabraded and belong to particles that have been collected originally by water holding silica in solution. By washing in acids the colouring matter of the particles may be entirely removed, and then it is seen that they are imperfect cyrstals. But the cementing matter is not always ferruginous; a felspathic cement holds them together with *used* mica evidently derivative, and sometimes with graphite.

Another variation in character of the Hawkesbury rocks is in their cohesion. In 1850 I was Chairman of the Artesian Well Board, and remember the difficulty we had in procuring tools hard enough to pierce the quartzose sandstone at the gaol in Sydney; the boring after a small depth was abandoned—one of the workmen precipitating the conclusion by blocking the bore-hole. But in parts of the Railway lines, there have been instances, as stated to me by the Engineer-in-Chief, when the largest blocks have been shivered to atoms by a not very heavy fall over an embankment.

The distinguishing features of the Wianamatta beds compared with the generally level horizon of the grits, sandstones and conglomerates of the Hawkesbury rocks are their greater proportion of calcareous matter; and in the region of the shales, the smooth rolling surface of the country. In the creeks formed by the synclinal slopes of the land, the Hawkesbury sandstones, much water-worn, are seen to underlie the Wianamatta beds.

Victorian Palæontologists claim for that Colony the existence of a coal formation of the same age as the Wianamatta, and I have myself long ago pointed out that certain beds at the Barrabool Hills resemble very closely certain strata about Camden, in New South Wales. But if the latter are proved to be of younger age than that which has been assumed for them, it is not necessary to place the two series (so widely separate in space) on the same actual horizon.

We have not recognised in New South Wales the Cycadeous plants of Victoria, nor is there a perfect agreement in the phytology of the Wianamatta and Victorian strata. In 1861 I mentioned ("*Recent Geological Discoveries, &c.,*" p. 45) three of M'Coy's New South Wales Plants, *Gleichenites odontopteroides* (called Pecopteris by Morris and Carruthers); *Odontopteris microphylla*, and *Pecopteris tenuifolia*, as occuring in the Wianamatta beds; these are not reported from Victoria, whilst *Sphenopteris alata, Brong. (Grandini of Goepp. and Schimper)* from Newcastle, belongs to the Old Carboniferous in Germany, and not to any Mesozoic formation.

In the list given in "Progress Report of Victoria, 1874," Professor M'Coy mentions three species of *Gangamopteris*, from his Upper Carbonaceous beds; 2 Neuropteris, 1 Pecopteris, 3 Sphenopteris, 1 Tæniopteris, with 3 Zamites and 1 Phyllotheca from the Lower Carbonaceous and only one animal form, *Unio Dacombii.* The alleged abundance and value of coal in these beds have been proved to be a myth. There is, however, more coal therein than in the smaller area of the Wianamatta and Hawkesbury rocks; and probably that is the reason why the Professor would place them *below* the former group of New South Wales. But when we consider the great improbability that a series of strata having a thickness of at least 5,000 feet could ever have existed between the Hawkesbury and Wianamatta series, and that not a trace remains anywhere in New South Wales of such interpolation,—that the fossil evidence is in opposition to it,—and that the areas are totally disproportionate,—it would appear a mere caprice of fancy to hold such a notion as that hinted at.

It may be well to make a final remark respecting Mr. Brough Smyth's idea that the coal beds of New South Wales lie on "*limestones.*" (*Progress Report, p. 26.*) Had he visited them himself he would have seen that limestone, as such, is rather a rare rock in connection with the New South Wales deposits of coal, which clearly interpolates the marine beds, but the latter are more frequently conglomerates, or sandstones and grits. The Upper coal measures rest frequently on granite and slates as well as on other rocks. The limestones in the Carboniferous rocks are rare, being few and of limited extent and far between. The author just mentioned considers the relation of the "*coal-bearing*" to "*palæozoic rocks*" as "obscure," but it is not obscure to those who have examined for themselves, nor more so than the feeling which induces *philosophers* to keep out of sight and ignore the evidence which contradicts their own pre-conceived opinions.

Mr. Charles Moore (of Bath) F.G.S., enumerates 171 species of Secondary animal fossils from Queensland, all sent to him for description by myself; and sixty-two from Western Australia, of which twenty species are common to England and that Colony.

The latter collection belongs chiefly to the Lower Oolites, Upper and Middle Lias; and the former embraces the Upper Oolites and Cretaceous formations. Mr. Brown, Government geologist in Western Australia (Report of 1873) mentions Mesozoic beds in the Darling Range, and, again, on the South Coast, from Cape Rich to beyond Mount Barren, and as far as Cape Esperance. Saliferous and reddish sandstones, &c., are the chief rocks. On his chart they and their detritus occupy seven degrees of latitude, and from one to three of longitude. But there is nothing defined as to fossiliferous evidence, except about Champion Bay. From Wizard Peak and Mount Fairfax I have received numerous fossils through the agency and kindness of the Hon. F. P. Barlee, F.R.G.S., Colonial Secretary, and the Rev. C. G. Nicholay, of Geraldton, who not only added to my collection, but supplied me with a personal survey of his neighbourhood on an enlarged scale, and with more minute details than Mr. Brown's chart exhibits.

There does not appear to be any fossiliferous evidence of Mesozoic formations in South Australia, where the rocks are chiefly Palæozoic, Metamorphic or transmuted and Tertiary.

In Tasmania, there is, no doubt, about the same evidence as for New South Wales. Victorian geologists believe that the coal of Jerusalem is Secondary. I was inclined to think that the neighbourhood of Green Ponds and Bagdad betrays a resemblance to some portions of the Wianamatta shales and sandstones of New South Wales. But the area there is far from extensive.

Mr. Gould, who surveyed considerable portions of the Colony, says nothing leading to the idea of any extensive Secondary areas; and whatever hold they may have on the mind of a geologist who has not carefully observed, must owe it to preconceived notions as to the age of the coal, which has of late established its Palæozoic character as unmistakeably as the seams of Anvil Creek, &c.

Coal has been reached on the Mersey under the marine fossiliferous beds, as I always held it would be, in spite of vaticinations to the contrary.

Passing over to New Caledonia, the Secondary formations are represented by Triassic, Liassic, and Neocomian rocks or fossils.

On the 6th July, 1863, a paper by M. Eugene Deslongchamps was read before the Linnean Society of Normandy on the Geology of Hugon Island, in New Caledonia, in which mention is made of a Pecten and fish scale from Cape St. Vincent, on the S.S.W.

Coast, collected by M. E. Deplanches. Millions of an Avicula (*Monotis*), allied to *M. salinaria* of Goldfuss, of which M. *Richmondiana*, of Zittel, is a variety, also occur. Astarte, Turbo *Jouani*, and one other; Spirifera *Caledonica*; S. *Planchesi*; Scyphia *armata*—all these are Upper Triassic.

M. Garnier's fossils, examined by M. Fischer, were pronounced to be *Monotis*; *Halobia* (an Austrian species), and Mytilus *problematicus* of the same formation.

The supposed Jurassic rocks contain Nucula near N. *Hammeri* (De Fr.) a Littorina, Cardium, and an Astarte resembling A. *Voltzii* (Goldf.) M. Fischer believes, however, that these are more likely to be Triassic also.

M. Munier-Chalmas names also as Jurassic, Ostrea *sublamellosa*; Astarte (or Tæniadon) *præcursor*; Pellatia *Garnieri* and Cardium *Caledonicum*.

A large *Pinna* seems to represent the Cretaceous rocks. A tolerably full account of the Geology of New Caledonia will be found in my Anniversary Address to the Royal Society of New South Wales on the 12th May, 1875.

New Zealand exhibits abundance of proofs that Secondary formations exist there, and not the least remarkable fact is that Professor Hochstetter in 1859 discovered there the same *Avicula Richmondiana* as above, and Halobia *Lomelli*, Avicula *salinaria* with Monotis, Spirigera, Spirifera, &c., belonging to the Triassic epoch.

In my paper "*On Recent Geological Discoveries*" I collected as much of this kind of information as I then could; but since then the skill and labour of the Geological Survey of New Zealand, under the direction of Dr. Hector, have produced an abundant harvest of scientific details; and to the able publications and Reports from that authority I may refer those who are interested in the development of that most interesting group of islands. They will find there ample evidence as to the existence of Triassic, Jurassic, and Cretaceous, as well as of Palæozoic rocks. The Saurian discoveries of Mr. T. Hood Cockburn Hood also deserve commemoration; nor must the labours and great discoveries of Dr. Haast be unremembered.

So far as the Trias is concerned, Hochstetter's discoveries of the genera and species about Richmond have been rivalled by Captain Hutton in Southland, Otago, who found in 1872, on the Moonlight Range, *Monotis Richmondiana* (Zitt), and *Halobia Lomelli* (Wissm). On the western slope of Hokanuis, and on the south side of the Wairaka Hills, he obtained the same species, with others, proving that the rocks are the same as the sandstones of Richmond, near Nelson, and also the Triassic age of the deposits. (*Geology of Southland. Report of Explorations, Geol. Surv., N.Z., p.* 104.)

Not very distant the same careful observer detected some of the same species as occur in Queensland in the middle Jurassic formation, described by Mr. Moore, *e.g.*, *Astarte wollumbillaensis*, with other genera and species, that link in the South with the North Island (p. 105). These discoveries justify the inference that Triassic rocks *are* probably present also in New South Wales.

When I first announced in 1860 the proof that Secondary fossils did exist in Australia, exhibited in Sydney, and forwarded to Sir Henry Barkly for Professor M'Coy's inspection I especially mentioned the occurrence of Cretaceous species. This was doubted, and the whole series classified as "*not higher*" than the "*lower part of the great Oolite.*" But in 1866, the Professor himself announced from another part of Queensland the occurrence of two *Inocerami*, and two *Ammonites*, from the Flinder's River district. He also announced an *Icthyosaurus*, a *Plesiosaurus*, and a *Belemnitella*, from lower Cretaceous strata of the same district.

Mr. Moore says, of the Wollumbilla fossils, "that they all belong to the *Upper Oolite* may with safety be inferred, but the Cretaceous beds have a claim to be considered, and he established the existence of the genus Crioceras, which was first reported by me.

In 1872, Mr. Daintree, F.G.S., read his Notes on Queensland, before the Geological Society, the marine fossils illustrating which were (as before stated), described by Mr. Etheridge, F.R.S., F.G.S., Palæontologist to the Geological Survey of Great Britain. The number of Oolitic species recorded is six, and of Cretaceous, twenty-five.

The expedition of 1872, in the Cape York Peninsula, in which Mr. Norman Taylor, of the Victorian Survey, was Geologist, has added to the list of Secondary fossils in Queensland. These were sent to me for inspection by the Minister for Public Works in that Colony, and at his request forwarded to the Agent General in London. They have not yet been fully described.

A still further amount of Cretaceous fossils forwarded by Mr. Hann, the leader of the Expedition of 1872, to Mr. Etheridge, and a large collection in my own cabinet, remain yet to be determined.

This is sufficient to show the extent of Mesozoic formations developed since 1860.

Mr. Daintree reckons the areas of the Cretaceous and Oolitic formations in Queensland at 200,000 square miles; the Carbonaceous (Mesozoic) at 10,000, and the Palæozoic Carboniferous at 14,000, whilst the Devonian and Upper Silurian occupy 40,000. The two younger, therefore, are more than *five* times as extensive as the older.

After the Norman Taylor collection had gone to England, I received three or four specimens from the Table Mountain, between Hann's Camps 41 and 42 (*Northern Expedition Report*) and forwarded them to the Queensland Agent General in London, for inspection by Palæontologists at Home. Mr. Etheridge, the Palæontologist of the Survey of Great Britain, considers the fossils in that conglomerate rock to be a species of *Hinnites* like H. *velatris* and an *Ostrea* like O. *Sowerbyii*, and that they belong to the Oolitic series. The same conglomerate as I learn by a more recent arrival, occurs on the high ranges between the Palmer and Cooktown, under the deposit which Mr. Daintree calls Desert sandstone. It is a coarse rock containing broken shells in a sandstone full of partly rounded pebbles. Mr. Etheridge also considers the Walsh River series to be of Lower Cretaceous forms. Some specimens of plants supposed to be Glossopteris were also forwarded by me to Europe, with the shelly rock. Mr. Carruthers' determination is, that they were not of that genus, but rather a form of Tæniopteris nearly allied to *Stangerites ensis* (Oldham and Morris in the Indian Survey Memoirs), which Schimper calls *Angiopteridensis.* Another specimen which I did not see in the great collection, but of which I had a drawing from Mr. Taylor, was considered by several geologists in Queensland, &c., to be Orthoceras, and, therefore, Palæozoic. Mr. Daintree says there were several specimens *like Orthoceras;* and so I think the one in question was, but I considered at the time that there was no Orthoceras present in the box, but a good many Belemnites, and I considered the sketch referred to was of the same genus.

I have since received the following statement,—"There was no specimen of Orthoceras in the entire series."

I have also received a list of the genera of Walsh River fossils, in Mr. Etheridge's handwriting. It is as follows, making all of them lower Cretaceous:—

Ammonites, allied to A. Clypeiformis.
Ammonites sp.
Crioceri.
Belemnites.
Myacites.
Byssoarca.
Solemya or Iridina.
Arca.
Panopæa.
Inoceramus.
Hinnites or Avicula.
Cytherea.
Cyprina.
Myoconcha.
Pecten.
Teredo or Teredina, in fossil wood.

An opinion has been adopted that the Mesozoic fossils from Queensland, both those described by Mr. Moore and these by Mr. Etheridge, were in mere drifted nodules. Mr. Taylor assures me that such is not the case with the latter, and I long ago gave a section of the beds at Wollumbilla, proving as in the York Peninsula, that the nodular masses were derived from a soft shale, being in fact concretions. If they have been drifted they have not travelled far.

Mr. Taylor (*Hann's Report*, p. 13) seems to have found the shelly deposit before mentioned on "a flat-topped Carboniferous range," (on 9 Sept., 1872), and by a report of April, 1875, from Cook Town, it appears that a fine seam of bituminous coal has been discovered at the junction of Oaky Creek and the Endeavour River, 20 miles from Cook Town; but from the determination of Mr. Carruthers, this coal (confirming, however, Mr. Taylor's statement) is not of the Glossopteris age. The coal of the latter series is not known to extend further north than 20° 35′ south.

In Mr. Dalrymple's Report of his Exploration on the "North-east coast of Queensland" (*Brisbane*, 1873, p. 20.) that enterprising observer states that the flat-topped ranges and mountains about the Endeavour River have "*red* sandstone escarpments," a feature that assimilates it somewhat to the "New Red" or Triassic formation.

Tertiary Rocks.

Kainozoic of Duncan.

Throughout the whole of Eastern Australia, including New South Wales and Queensland, no Tertiary *marine* deposits have been discovered. There are, however, in various places of New South Wales patches of *plant deposits* which, according to the frequent notices of geologists, may be referred to some period of the Tertiary epoch. A silicified sandstone, or quartzite of this kind, full of impressions of ferns and leaves of trees, but not known to be now living, occurs at Jerrawa Creek not far from Yass. It is probably Miocene. On the summit of the Cordillera, near Nundle, above the Peel River Diggings, occurs a ferruginous bed full of leaves. On the Richmond River occurs a white magnesite, full of yellowish impressions of leaves. At Kewong, in the county of Gowan, there is a bluish deposit of fine aluminous matter with black impressions. From a depth of 60 feet in a shaft near Bungonia, a pale yellowish white deposit with similar impressions was brought up; and on the summit of a "made" hill, above Kiandra Gold Field, at a height of 4,000 feet above the sea, and in a region now partly covered with snow many months in the year, there is a deposit of black clay with such casts of leaves as occur in similar clay near Hyde in New Zealand.

In recent visits to various gold fields of the Western districts, I have found plant beds of somewhat similar kind either cut by the shafts or distributed in the wash-dirt below the alluvial deposits, underlying in some cases thick masses of basalt. Such occur at Gulgong; at Cargo; under Bald Hill at Hill End; and also at Blayney.

At Lucknow also occur deposits of branches and fragments of trees under the basalt, and on the Uralla Gold Field, and at Home Rule, on Cooyal Creek, lignite and woody matter of a similar kind were seen by me in the lowest deposit of the deepest shaft.

No botanist is willing to declare what is the exact age of such deposits; but some of the leaves are supposed to represent, among others, the foliage of *Fagus;* yet it was only in 1866 that a beech forest was discovered, by the Director of the Botanical Gardens, growing on the M'Leay River. On comparing the living leaves with the impressions in the various deposits mentioned I can see no specific identity. This want of identity indicates that however the plants may resemble living plants they cannot be of a recent period; and yet there are occasionally such close resemblances as to lead some good botanists to infer a recent period for some of them.

The most remarkable instance I have examined is on the coast, about 42 miles north of Cape Howe, where, at a place called Chouta (between Tura and Boonda), a cliff about 100 feet high, formed of sand and white silicate of alumina, contains beds of lignite charged with sulphide of iron, and which are full of phytolites much allied to the living vegetation. From the clays, some of which are nearly kaolin, articles of pottery have been formed. It has been proved that, by distillation, a fair proportion of lubricating oil may be produced from the lignitiferous clay, and other products are expected to result from these deposits. The cliff is about 60 feet thick from the sea to the top of the clays, and borings below the sea-level have shown a still greater thickness.

These deposits lie between the horns of the little bay at Tura and Boonda, resting at one end on the highly undulating Palæozoic rocks, and at the other on a mass of porphyry. They were, formerly, no doubt, deposited in a depression among the slopes of the hills, but the wearing away of the coast has left a cliff of clay and sand instead of the original cliff of hard rocks. It is remarkable that at the south end the rocks assume the character of a breccia of quartz cemented by siliceous matter (probably like a deposit mentioned by Mr. Gould as occurring in Tasmania) and in it analysis has detected the presence of gold, though some quartz veins at the north end contained none.

My impression at first was that the lignite is recent, but I place the deposits under the present head because it may be possible, notwithstanding the opinion of a botanical friend whose judgment is worthy of esteem, the plants are not recent. Baron Von Müeller, to whom I submitted them, hesitated to express an opinion. They are deposited in clays of various kinds, chiefly white. Some of the hardened clinker-like sands covering the clays remind me of the sands on the coast of Dorset, at Studland and Bourne mouth. If this be really a Tertiary locality, it does not contradict the general assertion at the commencement of this section, for no shells of any kind have been detected in any part of these beds. Swampy and stunted plants still grow on the sands, which are very wet, and probably reproduce the phenomena beneath them, with the exception of the white clays which were in part derived from the decomposed felspathic matter of the porphyry. In various parts of Manecro there are lignite-like local thin deposits, but on analysis they have proved valueless.

By far the most interesting discovery that has been made in relation to the plant beds, was realized in the basaltic district of the Forest between Orange and Carcoar.

The description of several new genera and species of fossil prints has been given in "*Observations on New Vegetable Fossils of the Auriferous Drifts*," by Baron F. Von Müeller, C.M.G., M.D., Ph. D., F.R.S., and L.S., Government Botanist, &c.; published by the "Mining Department" of Victoria, 1874. These have been discovered, not only in the Forest, but also in Victoria, at Haddon, Nintingbol, Tanjol, and at Beechworth. They seem to belong to the later Pliocene formation, and to consist of plants allied to the present forest-belt of Eastern Australia. An abstract of the first account of them was read before the Geological Society, on 22nd June, 1870, and afterwards copied from the Quarterly Journal (vol. 27) into the *Geological Magazine*, 1870, p. 390.

They consist of the following species, viz.—

Spondylostrobus	*Smythii*
Phymatocaryon	*Mackayii*
"	*angulare*
Trematocaryon	*McLellani*
Rhytidotheca	*Lynchii*
"	*pleioclinis*
Plesiocapparis	*prisca*
Celyphina	*McCoyi*
Odontocaryon	*MacGregorii*

K

Conchotheca	*rotundata*
"	*turgida*
Penteune	*Clarkei*
"	*brachyclinis*
"	*trachyclinis*
Dieune	*pluriovulata*
Platycoila	*Sullivani*
Rhytidocaryon	*Wilkinsonii*

and, probably, some others.

This last species was discovered somewhere to the west of Bathurst in one of the gold leads, in the beginning of March, 1875, on the 10th of which month I had the good fortune to rediscover it in the refuse from a shaft near Lumpy Swamp, in the Forest, between Orange and Carcoar, Baron Von Müeller having stated in his Report of 29th July, 1874, that we require to learn "what was the nature of their leaves and floral organs." In order to search for these, I made a second journey to the Forest, having first explored it in 1872, and found, together with four specimens of *Rhytidocaryon Wilkinsonii* and a number of already described species, several *leaves* embedded in a ligneous clay in the refuse of a shaft, together with portions of the branches of some tree or trees. The tissue of the leaves was in some cases so thin that it peeled off on touching. The collection, which included a few other specimens of seeds and seed vessels given to me by Mr. A. Montgomery, who lives in the neighbourhood, I sent on to the Baron, who has forwarded them to Professor Schimper, of Strasbourg, being unable at present to undertake their examination. In a short time, therefore, we may expect to know more about these interesting plants.

Professor M'Coy has enumerated in the list of Tertiary Victorian fossils between thirty and forty *Oligocene* species; thirty to fifty or more Miocene, together with many tropical types of Dicotyledonous plants; and from the auriferous drifts four Molluscs, six Marsupials, and a Dingo, with the wood and fruit of a Banksia and the foliage of Eucalyptus obliqua. These are partly *Pliocene* and partly *Post pliocene.*

The occurrence of Banksia (four species) in the Tertiary formation of Häring, in the Tyrol (see Clarke's "Southern Gold Fields," p. 173) and in Victoria is a highly instructive fact as to the ancient vegetation of the world. The seed-vessels of plants deep below the surface in the auriferous drifts of Victoria and New South Wales were also mentioned by me in 1860, in the work alluded to above (p. 173).

The thickness of the rocks in the Forest and at Lumpy Swamp vary somewhat, but an example or two will show the character of the country over the gold leads.

Alluvium	10 feet.
Hard basalt	40 „
Decomposing basalt	40 „

Washdirt.

2. At Tigeroo shaft, near which I procured the seed-vessels:

Earth	10 feet.
Basalt	85 „
Peat and shale...	10 „

Washdirt with seeds and leaves.

At Haddon, in Victoria, the fossil fruit was found in one shaft at the bottom of the following section, resting on Silurian slates. (See Lynch's plans. *Vegetable Fossils of Victoria.*)

Black soil	1½ feet.
Red clay	4 „
Lumpy red and black clay	26 „
Clayey honeycombed rock, with hard cores succeeded by zeolitic basalt	100 „
Do. decomposed at base	1½ „
Black clay	7 „
Drift gravel and sand (auriferous) Trees at the bottom	10 „
Auriferous wash dirt (Fossil fruits) ...	6 „
	156

At Beechworth (El Dorado) occur wood and leaves in variably coloured clay above coarse drift, covering black clay with wood and leaves; and below this, two to eight feet of washdirt, holding fruits and wood, resting on granite. (From Mr. Arrowsmith's plan. *Id.*)

Mr. Daintree has stated his views respecting the Desert sandstone of his map that it is a Kainozoic deposit, and once covered the greater part of Australia. In the places where it is in great force, in Northern Queensland it overlies the Cretaceous rocks, and underlies lava beds. It contains fossil wood; and a Tellina which I sent to Mr. Daintree, from the neighbourhood of Leichhardt's Crossing-place, on the Flinder's River, would, he says, if coming from the Desert sandstone, show that that formation is not lacustrine. In various parts of New South Wales there are cappings of fine hardened sandstone which may have some relation to the strata referred to.

Mr. Daintree has, however, mistaken the locality he gives to the Tellina. He received a portion of a *Trilobite*, and not a Tellina, from Barkly's Tableland, and a cast of a whole one, which would give to that locality a Devonian character.

Towards the north of the Cape York Peninsula the sandstones are barren of fossils, and about the Cape seem to have more the character of *Laterite*, resting on Porphyry.

Mr. Wilkinson, in his researches among the tin-mines of New England, recognized the drifts which in Victoria are considered Pliocene; and Mr. Norman Taylor and the late Professor Thomson, in their paper "*On the occurrence of Diamond near Mudgee*" (*Trans. Roy. Soc. of N.S.W.*, 1870, *p.* 94) make mention of older and newer Pliocene drift. Whether there be any fossil evidence for the propriety of these terms I know not. That there are drifts of different parts of one epoch I believe, and, perhaps, the divisions are good, even if the designations are too refined. Dr. Duncan has advised us to postpone the Lyellian designations for the present. Having very recently visited almost every locality mentioned in the paper, and examined for myself much of the alluvia of the Gold Fields in a large portion of the county of Phillip, I am prepared to testify to the extreme faithfulness of the description given by Messrs. Taylor and Thomson. My remark, therefore, about the term Pliocene is not to be taken as complaining of it, but as a justification for the introduction of some of the drifts in question under the present head. A distinction of time is however clearly marked in the character of the various deposits or in the difference of botanical remains.

Perhaps some of these deposits in the Gold Fields, as well as some of the shelly conglomerates at the mouth of the Flinders, had better be considered as belonging to the next division of my subject; and though placed as Tertiary, I am not satisfied they are such, as no positive proof exists by unmistakable evidence that they are so.

In the far Western interior, beyond the Darling, shelly deposits of fine sandstone have been reached in well-making, and by the kindness of my friend, Mr. Woore, C.C.L., of the Albert District, I have been just put in possession of several good specimens, together with fossil wood, apparently not very ancient, which I believe to be Tertiary.

There is no doubt a fine waterworn drift over large areas of the Auriferous and Stanniferous regions and in the southern part of Manero; but in many cases the drift betrays its origin, as the result of the disintegration of conglomerates, and such I believe to be the origin of the drift seen by Professor Liversidge near Wallerawang. (Report on Iron Ore and Coal Deposits, read before Royal Society, 9 Dec. 1874.) He compares it with the diamond drift at Bingera, alluding to the "nodules of conglomerate" in each; but this conglomerate may be found *in situ* in the coal-bearing beds close at hand.

Many drifts have undoubtedly been dispersed and re-agglomerated, and again dispersed from one age to another, and the fineness of the pebbles and their perfect attrition afford testimony as to their antiquity, though now called recent.

True Tertiary marine fossils occur on the south coast from Cape Howe to Cape Lewin, and have been described by Captain Sturt, Rev. Julian T. Woods, and Mr. Busk. They are also met with on the west coast as far as North-west Cape, in great abundance.

New Zealand also contains a great number of Tertiary genera and species admirably detailed and arranged as belonging to the Upper Pliocene, Upper and Lower Miocene, and Upper Eocene, in a "Catalogue by Captain F. W. Hutton, F.G.S. (*Geological Survey, New Zealand*), Wellington, 1873, of Tertiary Mollusca and Echinodermata, in the collection of the Colonial Museum."

The classification is based on the *percentage* of recent species, the proportions of which are 76, 34, 23, and 9 *per cent*.

Quaternary Formation and Recent Accumulations.

The Quaternary Fauna of Australia has been so long known by the patient and skilful researches of Professor Owen, that there is no need to do more than refer to his writings, as the source of most of our knowledge respecting the strange animals that preceded the human epoch and perhaps extended into it. Huxley and others have also added to the general history of these creatures.*

Remains of reptiles have also been found both in New South Wales and in other parts of Australia, in quaternary deposits, as for instance, *Megalania prisca* (Owen), a Lacertian allied to the Varans and Lace Lizards of Australia, which had, probably, a length of 22 feet.

---

* An anecdote may be introduced here which may have some interest for visitors to the Australian Museum. In 1847, Mr. Turner sent to Sydney a box of bones from King's Creek, in Darling Downs, and Dr. Leichhardt, Mr. Wall (then Curator of the Museum), with myself examined them, and found there nearly the whole of the bones of the head, though in fragments only, besides other prominent portions of the Diprotodon skeleton, which had only been then partially known to Professor Owen, who had not at that time seen the *upper jaw*. So far, therefore, this individual was unique. With much trouble we put the bones together, and a cast was afterwards made of the skull, which is still in the Museum. A paper contributed by myself (dated 30th November 1847), and afterwards re-published in the Appendix to my Report of 14th October, 1853, ("*On the Geology of the Condamine River*"), and some letters from the late W. S. Macleay Esq., and Dr. Leichhardt, detailed the characters of the animal so far as they were then known, and the condition and other contents of Mr. Turner's collection. This would not deserve any mention here, but for the sake of introducing a curious event relating to the head of the Diprotodon alluded to. Mr. Turner sold his collection to the late Mr. Benjamin Boyd, who sent it to England. The ship was wrecked at Beachy Head, on the coast of Sussex, and the collection, forming part of the relics of the cargo which were sold, was taken to London, and Professor Owen bought it of the dealer who had become its owner, not knowing its history.

The Diprotodon appears not to have been limited to any one portion of Eastern Australia, for its remains have been found in South Australia and Queensland as far north as the York Peninsula.

In many of the "gold leads" also, fragments of bones are found. A section of one sample, at Wattle Flat, above the Turon River, is given in my paper on "Fossil Bones" (Q.J.G., S. xi. p. 405, 1855), and in Anniversary Address to Royal Society, N.S.W., 1873, p. 14."

In many parts of the existing region, all over the surface, wherever the basal rock is not denuded, as near Sydney, there are local deposits which might be called "till," were any Testacea found in them; and in the Interior there are widely spread accumulations of drift pebbles, which, as on the Hunter and Wollondilly, are rounded by attrition in their long journey from the mountains whence they have been derived. Sometimes, also, the breaking up of conglomerates has contributed to this drift.

On Peak Downs there are deep accumulations of drift, such as transmuted beds of the Carboniferous formation, igneous rocks, such as porphyry and basalt, and fragments of the older Palæozoic formation. Many of these are encrusted with thin calcareous cement, which forms cups of clear calc-spar in hollows of a fine porphyritic grit; the same grit occurring on the Warrego, on the Ballandoon and Narran ridges, with transmuted quartzite, also in wells there and on the Darling near Fort Bourke, in which drift fine gold was detected by me to exist on the Downs, and has been again reported to me from the base of Rankin's Ranges on the Darling River,—the furthest known western auriferous locality in New South Wales.

In 1869 I reported the discovery of the femur of a bird at the depth of 188 feet, in drift resting on granite, from a well in that part of Peak Downs (22° 40′ S.) which lies between Lord's Table mountain and the head of Theresa Creek, near the track from Clermont to Broad Sound. Compared with the bones of Dinornis in the Australian Museum, both the Curator of that Institution, and myself came to the same conclusion as to its genus, and accordingly it was reported in the *Geological Magazine*, as Dinornis. Professor Owen has, however, removed it into another genus *Dromornis* considering it to have belonged to a Struthioid bird. If it was such, of course (especially after the deep soundings between Australia and New Zealand, established by H.M.S. "Challenger" in 1874) the speculations I indulged on a possible former connection between those countries as illustrated by such a discovery are worth little. But if it was a *Dromornis*, then it falls in with the relationship to a present bird, the Emu, just as the Kangaroos of this epoch are related in structure to the gigantic Marsupials of a past age. But Mr.

Hood's discovery of Crocodilian remains in New Zealand seems to establish in another way some possible connection long ago with distant regions.

Crocodiles are yet common in Queensland. If the notion of a former connection of New Zealand with the latter region is rejected, we have a connection of another kind maintained by some geologists, and Australia is considered as forming a relick of a great Continent that formerly united what are now Africa and India with it. To this conclusion the existence of the plant deposits (discussed above) bears considerable testimony, and coupled with the wingless birds and crocodilian remains, an extension of the inference so as to include New Zealand is not unjustifiable. (See Mr. Blandford's paper "On the Plant-bearing Series of India, or the former existence of an Indo-Oceanic Continent," read before the Geological Society of London, 16th December, 1874.) Incidentally, that paper affords by its deductions, as to Permian times, an additional argument for the views I have expressed as to the epoch to which the Australian coal plants really belong being Palæozoic.

Looking to the Colony of New South Wales, we find that in more than one instance the present river channels have deepened since the drift first began to crowd their banks. I have traced one of these drift streams, sometimes at great heights above the valleys, for more than 80 miles. In other places I have found upon the surface, as Strzelecki did in other parts, minerals (especially ores of copper, tin, and lead) which were at a great distance from their sources; and in two instances that rare mineral, Molybdate of lead, of which no habitat has ever been yet found; and not more than a year ago a lump of Sulphuret of antimony, weighing three pounds, and exhibiting surface evidence of its being a drifted substance, was disinterred from the superficial ironstone gravel of an unfrequented place in the bush on one of the heights of the north shore of Port Jackson.

In the great plains of the interior bones of various gigantic marsupials, fishes and reptiles, are found bedded in black muddy trappean soil; and on Darling Downs, in Queensland, univalve and bivalve shells are found in some cases attached to the bones, or deposited over them in a regular series of layers, at intervals of several feet; and of these shells some are yet living in the water-holes of the creeks. These facts are generally known, but it was not till recently that the osseous relics have been found in different creeks throughout the whole of the slopes and plains at the base of the Cordillera in Eastern Australia; in Victoria, in South Australia, and in North Australia also. Of similar age are the accumulations of bones in caverns, as at Wellington; at Boree; near the head of the Colo River; at Yesseba, on the

Macleay River; at the head of the Coodradigbee; not far from the head of the Bogan, and in other places.

A magnificent collection of the remains in the Wellington Caves has been made, at the instigation of Professor Owen, at the cost of the New South Wales Government, with the superintendence of the Trustees of the Australian Museum, by one of them, the late Professor Thomson, and by Mr. Krefft, the Curator of that Institution.

The Reports of these gentlemen, together with more than a thousand partly determined specimens, were forwarded to Professor Owen, who has expressed his acknowledgment of the value of this collection, "as regards novelty, instructiveness, and encouragement for the future," and as an "important element in working out the ancient history of the forms of animal life peculiar to Australia."

The Coodradigbee caverns will repay research hereafter. It has already furnished me with bones of birds, in which those of an Emu are prominent.

The latter fact chimes in with the alleged *Dromornis* of Queensland.

Professor M'Coy has named bones of a Dingo in a cavern near Mount Macedon. If it be really a dog of this period in Australia, it is another link between the Quaternary and Recent times. Vicomte d'Archiac, however, doubts its antiquity: "*Rien*," he says, "*ne prouve que ce chien n'ait pas été introduit par les premiers hommes qui ont peuplé le continent Australien.*" (*Leçons sur la Faune Quaternaire*, *Paris*, 1866, *p.* 271.)

An expedition to Howe's Island made known in 1869 the existence of bones of birds and turtles embedded in the beach rock of the island. Afterwards, a collection of them was sent to me by Mr. Leggatt, of Fiji. I forwarded them to Professor Owen, who informed me that he was unable to determine to what they belonged, owing to their imperfect state; but they undoubtedly belong to some period near to the present, as the rock is a coral limestone, common to the coasts of the Pacific Islands; and that deposit also contains a Bulimus scarcely distinguishable from a living shell of the same genus off the Island, and eggs of Turtle also embedded as in Raine Island in the Barrier Reef. (*See Trans.*, *Roy. Soc.*, *N.S.W.*, 1870, p. 37).

Within the last few years, the drifts of the Cudgegong and Macquarie Rivers have been searched for diamonds, first reported in 1860 by myself as occurring in numbers in the latter river. Many thousand examples have been found, but they are chiefly small and of little value; though a few have been found of larger size, and have been cut and polished.

A few others have been brought to me from other localities in New South Wales, and a few also have been found in Victoria.

In other publications I have treated of them; and since then, the Bingera Diamond Field has received careful attention from Professor Liversidge who has described its condition accurately.

Those found since 1860 have fully justified the heading of my notice published that year ("Southern Gold Fields," p. 272),—"NEW SOUTH WALES A DIAMOND COUNTRY."

Some years since I reported on the occurrence of mercury in this Colony; but my expectation of the discovery of a lode of Cinnabar has been disappointed. The Cinnabar occurs on the Cudgegong in drift lumps and pebbles, and is probably the result of springs, as in California. In New Zealand and in the neighbourhood of the Clarke River, North Queensland, the same ore occurs in a similar way. About 1841 I received the first sample of quicksilver from the neighbourhood of the locality on Carwell Creek, on the Cudgegong, where the cinnabar is found. I proposed a full examination of that locality when I was in the neighbourhood in February, 1875; but the state of the weather was such as to preclude the possibility of doing so during my limited stay. But I was informed that the progress of the mine was satisfactory.

As connected with the drifts may be mentioned the occurrence of gems of all kinds in all the rivers where auriferous deposits occur, and subsequent years have only served to abundantly confirm my statement of 1860 as to the general distribution of them in the gold-bearing districts.

In examining the gold alluvia at a variety of shafts about Gulgong, Home Rule, and other places in the county of Phillip, I was struck by three prominent circumstances which have bearings upon the present and future of that region.

1. No shaft is, so far as I learned, deeper than 200 feet.
2. The gravels of the alluvia were composed of pebbles and fragments of rock common in the vicinity—derived from Carboniferous and underlying strata, with occasional fossils.
3. The quartz pebbles were in some cases perfectly rounded, in others the quartz was in fragmentary lumps, as if recently broken from reefs. These did not appear to occur together.

The conclusion I drew from the latter fact was that two periods of destruction and one of abrasion of underlying reefs had taken place at an early period of alluvial deposition. A fourth circumstance might be commented on. In the deposits of the shafts a multitude of well worn abraded lumps of jasper, silicified fossil wood, and semi-opal of various tints and chalcedonic interchanges, in some instances themselves decomposing, so as to exhibit the fibres of the wood from which they had been formed by transmutation, arrested attention, and showed that either an older series of Carboniferous rocks had suffered such changes, or the beds of the series which now exhibits itself in various outliers had undergone the process. My friend Mr. Lowe, of Gooree, has made a most extensive collection of these altered fragments, in which are many most beautiful specimens. It will probably never be rivalled, as he collected them from time to time as they were disinterred by the diggers. A great number also were coated with a shining transparent envelope of what I believe to be a deposit of silicified water. Elsewhere (Trans. Roy. Soc., 1870, p. 11) I have dwelt upon this; and it also attracted the attention of Professor Thomson and Mr. Norman Taylor. These deposits are frequently covered by a great thickness of basalt, upon which frequently lies a more recent drift partly derived from older drifts. The colours of the alluvia, now long exposed, rival in some degree those poikilitic hues which distinguish the west end of the Isle of Wight.

A drift of local kind also occurs over large areas in Maneero in the neighbourhood of the auriferous strata, as also in New England over the country of the tin mines, which exhibits the same sort of alluvia as the gold fields, and in which also gold occurs. In 1851-3, when I first discovered Tin in the Colony, it was generally in association with gold and gems. Messrs. Ulrich, Wilkinson, and Liversidge have since that time made local explorations both in the alluvia and in the beds from which they have been derived. There are deposits of opals besides those in the gold drifts; and on Lawson's Creek a feeder of the Cudgegong agate breccias and opals occur. Opaline veins also occur in the basin of the Abercrombie River, and in that of the Victoria of Queensland.

At the mouths of the Richmond and Clarence Rivers gold is found distributed in the sands and covering pebbles of the sea beach; a similar distribution is found in the sands of Shell Harbour (where the accumulations abovenamed occur) and some gold was extracted. Other spots give similar indications; and one specimen of gold was brought up from the sea bottom by the sounding operations of H.M.S. "Herald," off Port Macquarie. Gilded pebbles also occur on the west coast of New Zealand.

Numerous instances have also been recorded of gold having been found in the gizzards of wild fowl and of domestic poultry, in various parts of the Colony, confirming, with the above-mentioned facts, the almost universal distribution of the precious metal in river-drifts and superficial deposits. Some of the abovenamed examples of gold collected by birds were exhibited by me at Sydney and in Paris in 1855, and are still in my possession.

All along the coast, from Torres Strait to Bass's Strait drift pumice may be found wherever there is a lodgment, generally in the north corner of the little shore bays. That this has gone on for ages is apparent, as in one part of the coast north of Wollongong there is an accumulation of water-worn pumice, some distance from the shore and beyond the reach of the present waves. It is supposed to come in during easterly gales, from the volcanic islands to the north-east. In 1841 this fact, and all the evidence then collected in relation to such drift and "atmospheric deposits of dust and ashes," were published in a paper I forwarded to the *Tasmanian Journal*, of which D'Archiac (*Prog. de la Géol.*) was pleased to say it contained all that was known on the subject.

Subsequently received facts have only confirmed what was then stated.

Along the coast of New South Wales are found ranges of Dunes, with a variety of shells, some of them rare, others common, as on Port Hacking and Cronulla Beach; along the shores of Botany Bay; on the great flat between the Hunter and Port Stephens, and along the Macleay River, which now passes for many miles through the shelly accumulations; and about Moreton Bay and in more northern coast openings, shells and marine refuse form deep deposits, from which, as in Illawarra and Broken Bay, a considerable profit is obtained by dredgers and shell-collectors, for the production of lime.

Raised beaches also occur at various heights on rocky projections of the coast, indicating elevation of the land, of which there is distinct evidence in the recent period, not only in Moreton Bay, but near Sydney and thence to Bass's Strait; also on both sides of that Strait, and as far as Adelaide and King George's Sound. Mr. Selwyn gives data for assuming the elevation of the land to have reached occasionally 4,000 feet in Victoria, but he has no evidence of Tertiary marine fossils above 600 feet. Unfortunately, on the eastern coast, having no marine Tertiaries, we have to found our deductions, as respects New South Wales, on less secure data. Yet we have here evidence of another kind and pot-holed surfaces of considerable extent have been found by me at various heights from 300 to nearly 3,000 feet.

In a brief Memoir like the present it is impossible to quote all the authorities, nor has time allowed a more satisfactory digest or a wider range of statements. What has been thus collected is brought together in the design of giving a concise summary of the general Geology of the Colony, omitting, on account of its perplexity, all specific reference to the igneous rocks traversing, covering, transmuting, or supporting the Sedimentary deposits.

In *this* Edition many new facts have been introduced with the view of bringing on the discoveries that have been made from time to time to the present period, when a new system of geological inquiry has been just instituted in this Colony.

If private independent travel and research have not been able to accomplish more than this abstract discloses, it may be hoped that now the Government has made up its mind to undertake the work from its own resources, pecuniary and official, more will be accomplished than has hitherto been done to work out the intricacies of Australian geology, to accomplish which in minute and thorough detail, will probably require the united exertions of many a worker in the field and the cabinet to the middle of the next century at least. In the preceding pages it has been my lot to mention many of my own discoveries; but it has not been with any desire to overrate my endeavours or exertions; and some I have altogether omitted.

In the *first* Edition of this paper mention was merely made of the Cape York Peninsula, where ferruginous deposits occur on the lower slopes and bases of porphyry hills. I may repeat here what was added in the *second* Edition. Those deposits were examined at the Mint, and no gold was detected; but on a recent comparison of their lithological character with that of Tertiary beds from Flemington (in Victoria), I believe them to be, if not Tertiary, of similar origin to the *laterite* of India, and of the Islands in the intermediate sea.

Dr. Rattray, of H.M.S. "Salamander," who furnished me with a map, and a collection to illustrate it, from the neighbourhood of Cape York, and whose paper was read by me, in his absence, before the Royal Society of New South Wales, more recently published his views *in extenso* before the Geological Society of London. He therein attributes to me an opinion that the thick sandstones of the Peninsula are of the age of the Hawkesbury rocks of New South Wales.

I do not remember that I have expressed any opinion on this sandstone; what was submitted to me was considered by me far younger. That such sandstone, and even older deposits between Cape York and the Gilbert River, may exist in the interior of the Peninsula, is far from improbable. The data at present are insufficient for further comment. It may belong to the Desert sandstone of Daintree.

But this inference may be permitted that, as Cape York is so short a distance from the gold-bearing deposits of New Guinea, and as is now proved, all the rivers running to the Gulf of Carpentaria, from the Mitchell to the Nicholson inclusive, rise in auriferous ranges, gold will probably be found in some parts of the country, along the back-bone of the Peninsula; and although my past examination of the rocks in the Louisiade Archipelago has not proved gold to exist there, yet I agree with Mr. Daintree, in his last Report to the Queensland Government, that the strike of the older formations justifies the belief that that Archipelago, and, I may add, other portions of the lands insulated in that part of the Pacific, will eventually furnish their quota of the precious metal.

Several collections of New Guinea rocks have been sent to me; but although it was asserted strenuously that gold was found in them in the district visited by H.M.S. "Basilisk," I have not been able to recognise the existence of any auriferous matrix, though it is well-known that alluvial gold was discovered during the visit of H.M.S. "Rattlesnake" on the coast at the other side of the Island. I find, however, that nodules of excellent hæmatite occur at New Harbour about 100 feet above the sea. We may hope for satisfactory additions to our knowledge of that great Island from the results of the Expedition so nobly undertaken by Mr. Macleay.

In 1870 I added a remark or two about the discovery of a living Ceratodus in the waters of Queensland in the preceding year; the only previously known existence of the genus being the *teeth* found in Triassic European rocks to which that name was given.

This was an interesting addition to the living Trigonia, the Cestracion, the Terebratula, &c., of Australia, which connect the present period with the forms of life once held to be extinct.

Inquiries respecting this curious fish have resulted in the discovery of other species than that first found (Ceratodus *Forsteri*) and what is more extraordinary, fosilized teeth, of which I was shown examples by Professor Wyville Thomson who found them in an excursion purposely undertaken in search of the fish during the stay of H.M.S. "Challenger" in Port Jackson.

Since the first description of the fish by Mr. Krefft, Dr. Günther, F.R.S., has published a valuable "Description of Ceratodus, a genus of Ganoid Fishes recently discovered in rivers of Queensland, Australia," in the Phil. Transactions (part II. 1871). The result is that both Agassiz and Pander had from teeth found in the Lias and Trias of Europe come to conclusions which the living Ceratodus fully justifies. Dr. Oldham also had reported Ceratodus teeth from Maledi, south of Nagpur, in India. Australia in this instance precedes India. The fish turns out to be allied to Lepidosiren, and its habits are amphibian, feeding on grasses and weeds in fresh water.

Dr. Günther goes into a most elaborate and minute examination of the anatomy of all parts of the fish and a comparison with other fishes of the same and different types. He sums up thus:—"The Dipnous type is represented in the Devonian and Carboniferous epochs by several genera (*Dipterus, Cheirodus, Conchodus, Phaneropleuron*), it is then lost down to the Trias and Lias, where the scanty remains of a distinct genus, *Ceratodus*, testify to its presence; no further trace of it has been found until the present period, where it re-appears in three genera, one of which is identical with that of the Mesozoic era. Now, at present, scarcely any zoologist will deny that there must have been a continuity of the Dipnoous type, and it is only a proof of the incompleteness of the palæontological record, that we have to derive all our information regarding it from only three so very distinct periods of existence. The *Dipnoi* offer the most remarkable example of persistence of organization, not in fishes only, but in vertebrates. On a former occasion I have shown that numerous recent species of fishes have survived from the period of the geological changes which resulted in the separation of the Atlantic and Pacific by the Central American Isthmus. In Ceratodus we have now found a *genus* which, as far as evidence goes, persisted unchanged from the Mesozoic era; and in the *Sirenidæ*, a *family*, the nearest ally of which lived in the Palæozoic epochs."

This is a most valuable link in the connection of the old geologic periods with the present era, and a fit conclusion for the account above given, however, unworthy that account may be, of Quaternary and Recent Accumulations.

No notice in this Memoir has been taken of igneous rocks; but it may be suitable to state that there is in all the various Sedimentary formations noticed distinct evidence of the presence of igneous action (*hydro-igneous* rather), and their transmutation through such and allied agencies has left an impress upon all the rocks more or less concerned.

No particular or special reference could enter into the object for which this Memoir is written; but it is to be understood that, though all the rocks have undergone a transmutation, this does not constitute what geologists have understood by "Metamorphic" system, of which, as before said, New South Wales, at least, shows little or no visible trace.

W.B.C.

2 June, 1875.

P.S.—In order to explain the position of Glossopteris in the Paleozoic marine deposits, I have appended two vertical sections, one, by myself, previously published in the "Transactions of the Royal Society of Victoria, 1861," illustrating the coal seams at Stony Creek; and the other showing the deposits at Greta, near Anvil Creek, which has been reduced from one on a larger scale, kindly supplied to me by Mr. James Fletcher, Colliery Viewer, to whom I am also indebted for a collection of strata, the characteristics of which I have given after careful examination of them and of other specimens collected by myself on former occasions. The latter section illustrates a wide area on that part of the Hunter River. No. 2 is about 10 miles west of No. 1.

30th June, 1875.

[illegible]

[illegible]

[illegible]

[illegible]

[illegible]

[illegible]

[illegible]

[illegible]

[illegible]

[illegible]

[illegible]

[illegible]

[illegible]

[illegible]

[illegible]

[illegible]

[illegible] Coal.

[illegible]

[illegible] Very hard conglomerate.

[illegible] Ironstone band.

[illegible] Conglomerate.

[illegible] Coal.

[illegible] Blue rock, full of large [illegible]

[illegible] Boring. [illegible] conglomerate [illegible] to depth of [illegible] feet.

SECTION OF B. PIT, GRETA.

	Feet.	In.	[illegible]
[illegible]	[illegible]	[illegible]	[illegible]
Total	[illegible]	[illegible]	

SUPPLEMENTARY REPORT.

(By John Mackenzie, F.G.S., Examiner of Coal Fields, New South Wales.)

The Examiner of Coal Fields reports as follows:—

I now have the honor to submit to you my Supplementary Report, accompanied by ten vertical sections of the upper coal measures, a plan to a small scale showing the boundary and extent of the New South Wales Coal Fields, as far as I have examined it, with the places where the different sections were measured shown thereon. Two longitudinal sections, one for exhibition purposes, and the other to accompany this report, of provings across our lower coal measures by the Australian Agricultural Company, at Stroud, in the County of Gloucester, and six diagrams, to be lithographed for Exhibition purposes, showing the character, thickness, and portion mined out of the seams of coal worked at the different collieries; also four plans, showing the locality of the different collieries now at work in the Northern, Southern, and Western Districts.

Newcastle Harbour, and its facilities for shipment.

Newcastle, the trade of which is second only to that of Sydney, owes its great commercial importance to the different coal mines which have been opened out close to and within 32 miles of the harbour.

The Government coal wharf is 2,400 feet in length, and 24 feet 6 inches in width, and eight cranes are used for shipping the coal, three cranes, said to be capable of shipping 400 tons each in twelve hours, one of 600 tons in twelve hours, and four of 1,000 tons each in twelve hours. Or 7,600 tons of coal can be shipped in twelve hours into ships of the collier class, which do not require trimming, and alterations are now being made; four of the first mentioned and oldest cranes on the wharf are being removed, and replaced with three 15-ton cranes capable of shipping 1,000 tons each in twelve hours, and the distance between the cranes has been increased, thus admitting larger vessels to be berthed.

A line of wharf is now being constructed 8,000 feet in length, in 100 feet sections, 300 feet apart on the western side of the harbour, and on four of these sections four 15-ton hydraulic cranes, which are expected to be capable of shipping 1,000 tons each in twelve hours, will shortly arrive from England, and a railway to connect the wharf with the Great Northern Railway and collieries is now being constructed.

It is expected that in less than eighteen months coal will be shipped from this wharf, where vessels of the largest size can be loaded.

With a view of future extension for increasing the shipping appliances, the hydraulic pumping plant now arriving from England is sufficient to supply sixteen cranes, so that as the demand for our coal increases extra cranes can be erected at a short notice.

Between the southern end of this wharf and Bullock Island space is reserved for a 90-acre basin, around which a wharf a mile and a half long is proposed to be constructed, which it is intended to dredge to a depth of 23 feet, when vessels will be able to lie in slack water.

In addition to the Government cranes and staiths, there are the private shipping appliances of the Australian Agricultural Company, who have five staiths, capable of shipping 400 tons each in twelve hours, and the Waratah Colliery Company's shoots at Port Waratah, capable of shipping 600 tons in twelve hours.

Each of the Government cranes has a full and empty line of railway, and lifts the coal waggons of 6 to 10 tons, and slewing them round, discharges the coal into the hold of the ship.

The total quantity shipped last month was about 107,000 tons.

I shall now commence my observations on the Northern District Collieries, beginning with those in the neighbourhood of Newcastle, in the County of Northumberland.

PLAN No. I RE

Shewing the Position and Extent of the diff

IN THE

County of North

— NORTHERN D

REFERENCE

- **A** Australian Agricultural Colliery Company
- **B** Waratah Coal Company
- **C** New Lambton Colliery Company
- **D** Lambton Colliery Company
- **E** Co-operative Colliery
- **F** Newcastle Wallsend Colliery Company
- **G** Duckenfield and Minmi Collieries
- **H** Mess.rs Brown & Lamb
- **I** Hon. J Robertson & Others
- **J** Australasian Coal Company
- **K**
- **L** E C Merewether Esq.
- **M** Woodford Colliery
- **N** Four Mile Creek Collieries
- **O** Stony Creek
- **P** Anvil Creek Colliery
- **Q** Greta Coal & Shale Company

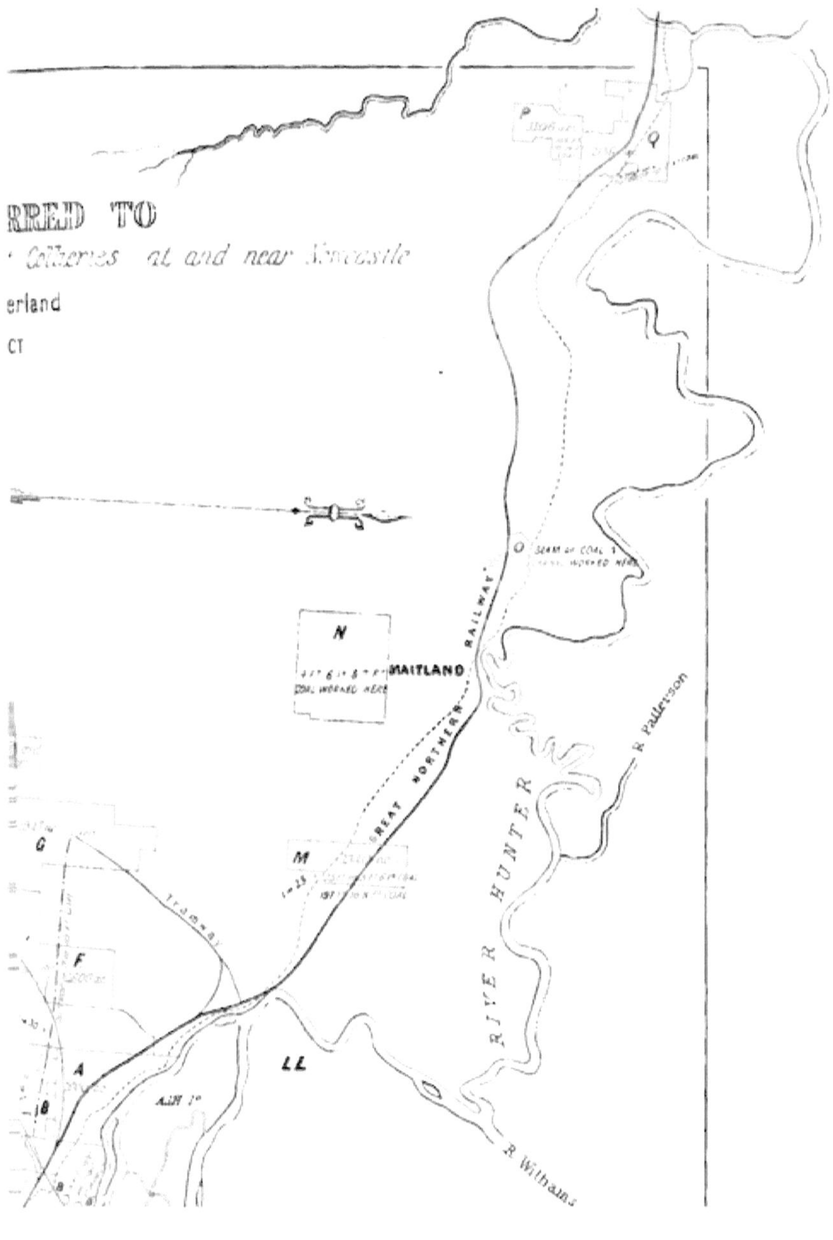

RRED TO
Collieries at and near Newcastle
erland
CT
P
Q
O
SEAM of COAL
N
MAITLAND
COAL WORKED HERE
RAILWAY
GREAT NORTHERN
R Patterson
RIVER HUNTER
G
M
Tramway
F
A
B
LL
R Williams

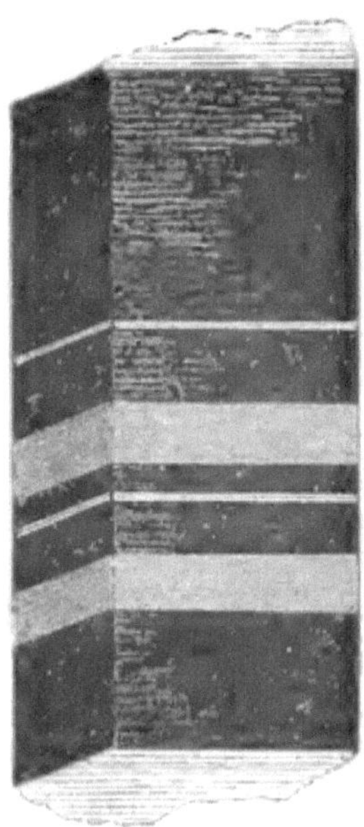

Shale and grey post, full of Glossopteris, Sphenopteris, and Phyllotheca.

	ft.	in.
Coal	4	0
Band	0	1
Coal	1	1
Coal, inferior (Morgan)	1	0
Coal	0	5
Stone band	0	1½
Coal	0	10
Coal and bands (Jerry Wag)	0	10
Coal	2	2
Total thickness ...	10	6½

Grey post, with Glossopteris, Phyllotheca, &c.

This section represents the seam of coal worked by the Australian Agricultural Company, at a depth of 200 feet, at their Hamilton pit, near Newcastle, from under their freehold land and that leased from E. C. Merewether, Esq. (see A on plan No. 1). It averages here 10 feet 6 inches in thickness, and is identical with that numbered 6 on the vertical section of the Newcastle Coal Measures. (See section A, No. 7.) It is very free from faults, lies very regular, and has an average dip of about 1 in 20 to the south-east. It is a free-burning bituminous coal, suitable for household, steam, smelting, gas, coking, and blacksmith purposes, and has an average specific gravity of about 1·23 to 1·29.

This Company has 4,000 acres of freehold land in this district, 720 acres of leasehold, and two pits at work raising coal from this seam—one by a 60 h.-p. horizontal engine, and the other by a 30 h.-p. horizontal engine—both engines being suitable for winding and pumping; two underground horizontal engines, for drawing loaded skips to the pit-bottom and working force-pumps—one 30 h.-p., and the other 15 h.-p.; two 14-ton locomotive engines and one 30-ton tank engine, for taking the coal from the pits to the Company's staiths, a distance of less than 2 miles, where ships of large tonnage are loaded. This Company raised 195,494 tons of coal, valued at £120,963, in 1874, and employed on an average

596 men and boys each day the colliery was at work. E. C. Merewether, Esq., is the General Superintendent, and J. B. Winship, Esq., the Colliery Manager.

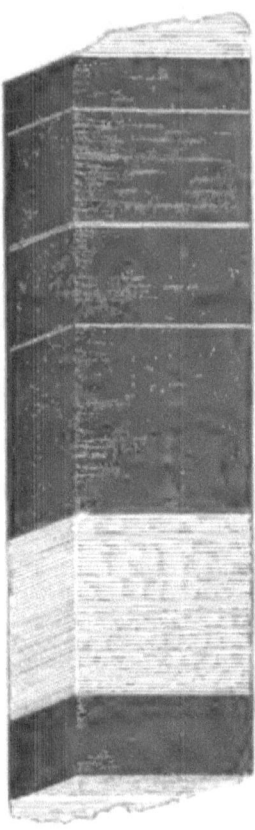

Shale and grey post, with Glossopteris, Phyllotheca, &c.

	ft.	in.
Coal	1	0
Band	0	0⅛
Coal	2	5
Stone band.............................	0	1
Coal	2	3
Stone band.............................	0	1
Coal	4	3
Total thickness ...	10	1¼

Fireclay, 4 ft.

Coal, 1ft. 8 in.

Grey post.

The above is a section of the seam of coal worked by the Waratah Coal Company, at Waratah, near Newcastle. (See B on plan No. 1, and No. 6 on vertical section, lettered A, section No. 7.) It is worked by adits driven into the hill side, where the coal averages about 10 feet 1 inch in thickness, 9 feet 1 inch of which is generally mined. It is a free-burning bituminous coal, suitable for household, steam, smelting, gas, coking, and blacksmith purposes, has an average specific gravity of 1·23 to 1·29, and lies very regular, dipping at the rate of about 1 in 30 to the south. They have 3,000 acres of land, five stationary engines, of an aggregate of 85 h.-power, and three locomotive engines, amounting to 130 h.-p., for taking the coal from the mine to the Newcastle harbour, a distance of about 4 miles; also 150 9-ton coal waggons, 4 miles of private railway, and a shipping place of their own, at Port Waratah, about 2½ miles from the Newcastle wharf. They raised 181,268 tons of coal, valued at £107,032, in 1874, and employed 285 miners and 106 men and boys daily when the colliery was at work, and 37 horses.

Mr. Moody is now engaged opening out an upper thick seam of coal on the southern portion of the property, similar to the one now being worked by the Australasia Company.

R. L. Moody, Esq., is the Colliery Manager.

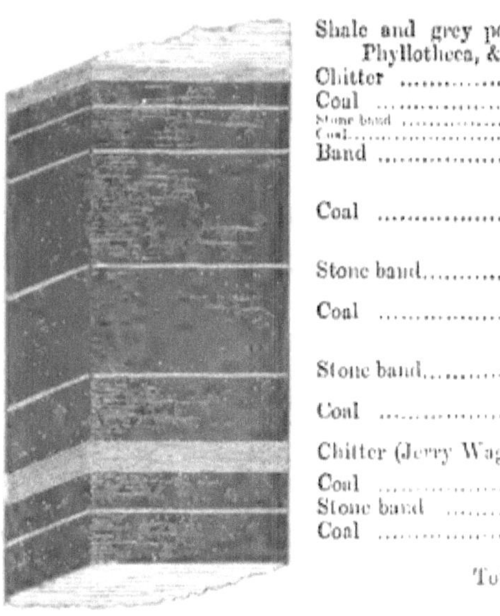

Shale and grey post, with Glossopteris, Phyllotheca, &c.	ft.	in.
Chitter	0	4½
Coal	0	4½
Stone band	0	0½
Coal	0	10
Band	0	0½
Coal	2	2½
Stone band	0	0½
Coal	2	1
Stone band	0	0½
Coal	1	3
Chitter (Jerry Wag)	0	7½
Coal	0	10
Stone band	0	1½
Coal	0	10½
Total thickness ...	9	9

Here we have a section of the seam of coal 9 feet 9 inches in thickness, 8 feet 1 inch of which is generally worked at the New Lambton Colliery, situated about 5 miles by rail from the Newcastle Harbour. (See C on plan No. 1, and No. 6 on vertical section A, section No. 7.) The coal is worked from a shaft 100 feet in depth; it lies very regular, with a dip of about 1 in 30 in a southerly direction; is a free-burning bituminous coal, suitable for household, steam, smelting, gas, coking, and blacksmith purposes, the specific gravity varying from 1·23 to 1·29. The property consists of 1,185 acres of freehold land; they have two engines of 20 and 16 h.-p., one of which raises the coal up the shaft, and the other is an underground engine used for hauling the coal to the bottom of the pit. The coal sent from this colliery in 1874 amounted to 133,805 tons, valued at £84,815 3s., and they employed daily when the colliery was at work 265 miners and labourers, and ninety-seven boys. This colliery is owned by Messrs. Alexander Brown and J. C. Dibbs.

Shale and grey post, with Glossopteris, Phyllotheca, &c.

	ft.	in.
Coal	0	10
Band	0	0½
Coal	2	4½
Band	0	0½
Coal	1	11
Stone band	0	0½
Coal	1	5
Chitter (Jerry Wag)	0	3
Coal	0	7
Penny band	0	0½
Coal	1	8½
Total thickness ...	9	2½

Fire-clay, 3 ft.
Grey post.

This denotes the thickness and character of the seam of coal worked at the Lambton Colliery near Newcastle, the property of the Scottish Australian Mining Company, situated within 5 miles of the Newcastle Harbour. (See D on plan No. 1.) Where measured it is 9 feet 2 inches in thickness, 8 feet 5 inches of which is worked, and is the seam of coal shown and numbered 6 on the section lettered A, section No. 7. It lies very regular, the dip being only about 1 in 30 to the south; a few trap dykes have been crossed, on either side of which the coal is "charred." It is a free-burning bituminous coal, suitable for household, steam, smelting, gas, coking, and blacksmith purposes, and has an average specific gravity of 1·23 to 1·29.

There are two main entrances to the mine—one by an adit and the other by a shaft 35 feet in depth, the coal being raised up this shaft by an engine at the rate of 750 tons of coal in nine hours.

They have an underground engine which draws the coal a distance of one mile from the winding shaft, and one of Cameron's Patent Pumping Engines for draining the mine, which pumps 500 gallons of water per minute from a depth of 220 feet, and Mr. Croudace informs me that it is the only engine of its kind in the Colony.

The ventilating furnace is one of the largest in New South Wales, having a fire surface of 20 feet by 10 feet, equal to 200 feet and is capable of being extended to 26 feet by 10 feet.

This colliery raised 127,768 tons of coal, valued at £80,188 16s. in 1874, and employed daily, when working, 224 miners forty-four labourers and others and thirty four boys; also twenty-one horses underground. Thomas Croudace, Esq., is the Colliery Manager.

Shale and grey post, with Glossopteris, Phyllotheca, &c.

	ft.	in.
Coal	0	8
Band	0	0½
Coal	2	5
Band	0	0½
Coal	2	4
Band	0	0½
Coal	3	7½
Coal (inferior)	1	0
Total ...	10	2

Fire-clay and grey post.

This represents the seam of coal worked at the Co-operative Colliery, belonging to William Laidley, Esq., situated at Wallsend, a distance of about seven miles by rail from the Newcastle Harbour. (See E on plan No. 1.) Here this seam of coal is about 10 feet 2 inches in thickness, 8 feet 5 inches or thereabouts being worked, on an average. It is a free-burning, bituminous coal, used for household, steam, smelting, gas, coking, and blacksmith purposes, and has an average specific gravity of 1·23 to 1·29. This property consists of 1,500 acres, and the coal is brought to the surface by an adit or tunnel, near the entrance of which there is a stationary engine, which draws the coal in the miner's skips from the far end of the workings up a slight incline, and tips the coal into waggons waiting there to receive it. The coal is very regular in its thickness and quality, lies very flat, dipping about 1 in 30 to the south. A large quantity of coke is made here from the small coal by ovens erected close to the mine, which is equal to any made in England or elsewhere. The coal sent away from this colliery in 1874 amounted to 149,699 tons, valued at £94,982, and the average daily number of men and boys employed was 333.

James Fletcher, Esq., is the Colliery Manager.

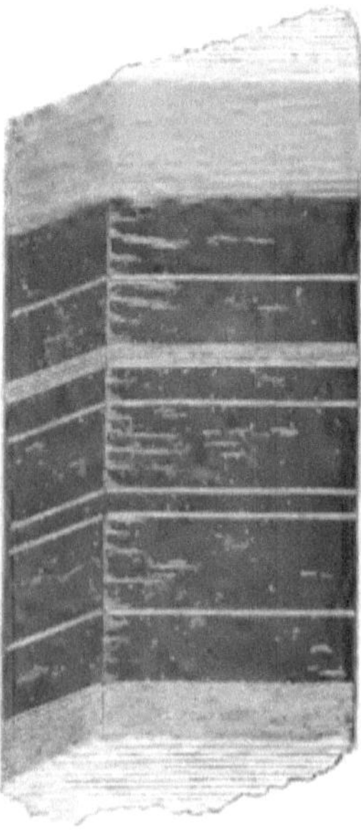

Grey post.

Blue shale, with Glossopteris, Sphenopteris, and Phyllotheca, &c.

	ft.	in.
Coal	1	2
Indurated clay	0	0½
Coal	1	0
Brassy coal	0	4
Coal	0	7
Stone band	0	0½
Coal	1	5
Stone band	0	0½
Coal	0	3
Band	0	0½
Coal	1	5
Coarse coal	0	2
Coal	1	0
Coarse coal, full of bands	0	11¼
Total thickness ...	8	5¼

This is a section of the seam of coal at the Newcastle Wallsend Colliery Company's C Pit, where it averages 8 feet 5 inches in thickness, 7 feet 6 inches of which is worked. (See section lettered B, section No. 8, and F, on plan No. 1.)

It is a free-burning bituminous coal, used for household, steam, smelting, gas, coking, and blacksmith purposes, and has an average specific gravity of 1·23 to 1·29, the mine being situated about nine miles from the Newcastle Harbour. There are several shafts sunk to this coal on the Company's 9,000 acres of land, but they are at the present time only using one for raising and winding coal. This shaft is 185 feet in depth, and 14 feet in diameter, and it has a 60 h.-p. engine, which lifted an average of 1,100 tons of coal per day of nine hours out of the mine in 1874.

The largest quantity raised in one day of nine hours was 1,270 tons, for which the manager deserves very great praise, as it would be impossible to raise such a large daily quantity at one shaft, unless everything above and below ground was well and systematically laid and carried out. They have two underground traction engines, one of 40 h.-p., and the other 30 h.-p.; one 15 h.-p. coal hopper-engine, and one 15 h.-p. engine for sinking shaft.

30 horses, 400 miners, and 200 labourers, wheelers, and mechanics, employed daily in 1874, when the pit was working.

They raised 240,000 tons of coal, valued at £163,000, in 1874.

The Colliery Manager is J. J. Neilson, Esq.

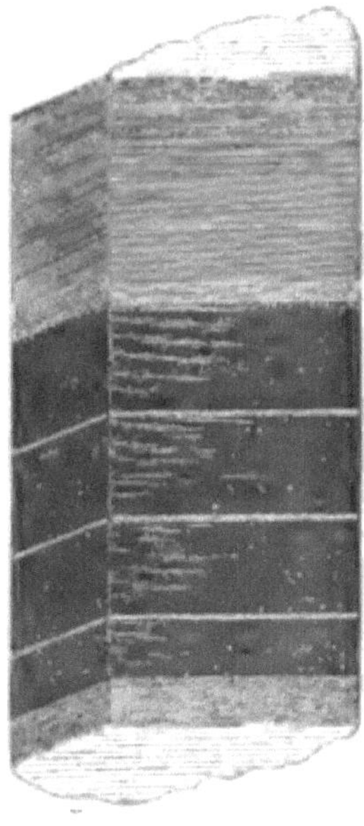

Blue shale, with Glossopteris, Phyllotheca, &c.

	ft.	in.
Coal and bands	3	6
Coal	1	8½
Band	0	0½
Coal	1	8
Band	0	0½
Coal	1	6
Band	0	1
Coal	0	9
Coarse coal..............................	0	9½
Total thickness...	10	1

This is a section of the seam of coal worked by Messrs. J. & A. Brown, at their Duckenfield Colliery, six miles by rail from the mine to shoots erected alongside the Hunter River, where vessels of large size are brought from Newcastle to be loaded with coal, a distance of about 12 miles. (See G on plan No. 1, and No. 6 on section lettered A, section No. 7.) The thickness of the seam is here about 10 feet 1 inch, but the upper part is so full of bands that they only work 5 feet 6 inches. It is a free-burning bituminous coal—specific gravity from 1·23 to 1·29—suitable for household, steam, smelting, gas, coking, and blacksmith purposes. This property consists of 2,767 acres, and the same seam of coal was formerly worked by Messrs. J. and A. Brown, a short distance south-east of the present workings.

The coal is worked from an adit driven into the hill-side by which it was first brought to the day in the latter part of 1874.

The quantity of coal raised in 1874 was 3,821 tons, valued at £2,594 4s., and the average number of miners, labourers, and boys employed daily when at work in 1874 was forty-four miners, forty-three labourers, and thirteen mechanics, and they are employing a very much greater number this year. They have two locomotives and three stationary engines of six, twenty, and forty horse-power.

Alex. Brown, Esq., is the Manager.

The above sections represent one and the same seam of coal, and the following fossil flora is found in the rocks and shales lying above and below it, viz.:—Sphenophyllum, Vertebraria, Clasteria, Gangamopteris, Sphenopteris, Pecopteris, Noeggerathia, Glossopteris, Phyllotheca, Anarthrocanna, with Coniferous trees and seed-vessels of Conifers, and fluted stems very like Calamites—with other plants not yet determined by any palæontologist. And the Rev. W. B. Clarke, M.A., F.G.S., &c., informs me that a palæozoic fish, "Urosthenes Australis," was found in one of the Australian Agricultural Company's pits, in a bed of grey grit lying a short distance over the coal now worked in this neighbourhood.

CATHERINE HILL BAY, LAKE MACQUARIE, AND REDHEAD DISTRICTS.

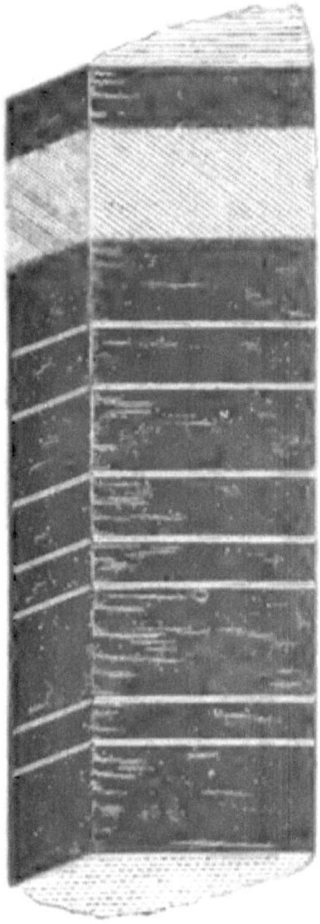

	ft.	in.
Grey post and conglomerate.		
Coal	1	0
Indurated clay, &c.	2	2
Coal	1	5
Band	0	0¼
Coal	1	0
Band	0	0½
Coal	1	6
Stone band	0	1¼
Coal	1	2
Band	0	0¼
Coal	0	10
Band	0	0¾
Coal	1	11
Band	0	0½
Coal	0	9
Band, runs out sometimes	0	0¼
Coal	2	0
Total thickness ...	14	0¼

This represents a section of the No. 2 seam of coal worked at the New Wallsend Company's Colliery, at Catherine Hill Bay, near Lake Macquarie. It is the No. 2, or second seam of coal in our Upper Coal Measures, is 14 feet in thickness, and the portion at present worked averages about 6 feet 8 inches, and dips 1 in 20 to 1 in 30 in a westerly direction. This is a new colliery opened out by Thomas Hale, Esq., and others, on 267 acres of land (see H on plan No. 1) fronting the Pacific Ocean and Catherine Hill Bay; it is bounded on the north by Messrs. Brown, Lamb, and others, and on the south and west by Messrs. Pope, Harding, and others, on which lands the same seam of coal is to be found.

The upper part of the seam is a splint coal, and the lower part a splint and bituminous coal; the specific gravity of some pieces I tested from the lower part gave 1·32. It is situated about 50 miles north of Sydney Harbour, and the coal is worked by an adit or heading driven into the seam of coal at a height of about 8 feet above the level of the sea, and a jetty has been constructed from the mouth or entrance to the seam of coal out into the bay, where steam colliers come and take away the coal from the waggons filled by the men hewing coal in the mine. A village called Cowper, consisting of about 150 people, now exists on these once desolate ranges, and a dispensary, hotel, post office, stores, &c., are erected thereon. The quantity of coal shipped from here in 1874 amounted to 18,147 tons, valued at £11,795 11s., and the average daily number of miners and labourers employed when at work was eighty, and six boys.—J. B. Winship, Esq., is the Colliery Viewer; and T. H. Hale, Esq., the Secretary.

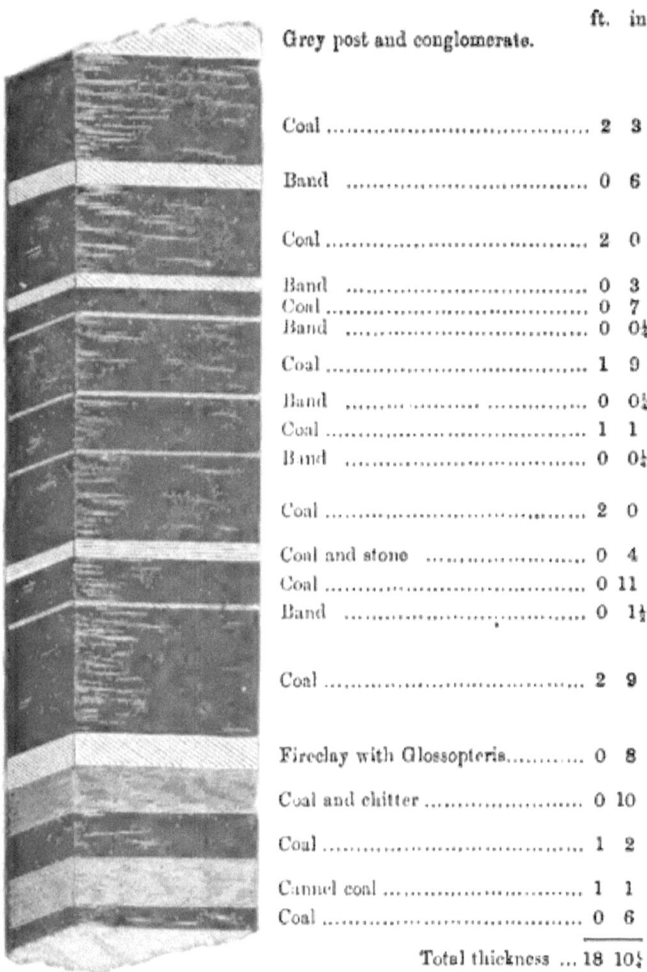

	ft.	in.
Grey post and conglomerate.		
Coal	2	3
Band	0	6
Coal	2	0
Band	0	3
Coal	0	7
Band	0	0½
Coal	1	9
Band	0	0½
Coal	1	1
Band	0	0½
Coal	2	0
Coal and stone	0	4
Coal	0	11
Band	0	1½
Coal	2	9
Fireclay with Glossopteris	0	8
Coal and chitter	0	10
Coal	1	2
Cannel coal	1	1
Coal	0	6
Total thickness	18	10½

This is a section of the seam of coal formerly worked by the Cardiff Company at Lake Macquarie, 18 feet 10 inches in thickness—10 feet 4 inches of which they worked. It now belongs to the Honorable John Robertson and others, and the coal is of very good quality, and I have no doubt that in a very few years we shall have this and the surrounding coal land, belonging to numerous other parties, connected by rail with the Newcastle Harbour. For the place where this measurement was made, see H on plan No. 1.

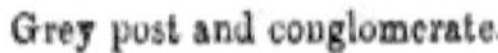

	ft.	in.
Grey post and conglomerate.		
Coal and bands of no value, about	12	0
Coal	0	9
Indurated clay	0	1
Coal	0	9
Indurated clay	0	5
Coal	1	5
Indurated clay	1	6
Coal	2	3
Band	0	0½
Coal	1	6
Band	0	0½
Coal	2	11
Fireclay 6 inches to	0	3
Coal	2	2
Total thickness ...	26	1
Conglomerate.		

This represents a section of the seam of coal lately opened out by a new Company called the Australasia Coal Company limited, whose head office is 30, Queen-street, Melbourne.

They hold 1,034 acres on mineral lease from the Crown, and 40 acres of purchased land. (See letter J on plan No. 1.)

The seam is identical with the one I have just described, it is here about 26 feet thick, the upper part being very full of bands; it is proposed only to work the lower 9 feet 2 inches, which is a free-burning bright bituminous coal, suitable for household, steam, smelting, gas, coking, and blacksmith purposes.

The specific gravity of specimens I tested was 1·213, and the dip appears to be about 1 in 30 west. It has an excellent roof and floor, and is very advantageously situated for cheap and economical working.

Mr. Moody, the Company's Manager, informs me that he has just completed a survey for a tramway 6 miles in length from the mine, to intersect the Great Northern Railway at a distance of 2 miles from the Newcastle Harbour, for which tenders are now being called. The lower seams of coal have not yet been proved on this property. J. P. Moody, Esq., is the Colliery Manager.

	ft.	in.
Grey post and conglomerate.		
Coal and bands	4	0
Indurated clay	0	1¼
Coal	0	3¾
Band	0	0½
Coal	1	4½
Indurated clay	0	5
Coal	1	0
Indurated clay	0	3
Coal	1	0
Coal	3	8
Indurated clay	0	2
Coal	0	6
Indurated clay	0	0¼
Coal	0	2
Indurated clay	0	0¼
Coal	1	3
Band	0	0¼
Coal	1	3
Total thickness...	15	7
Grey post.		

This denotes the seam of coal opened out by Messrs. Alcock and others (see K on plan No. 1) on 600 acres of land held by them at Redhead, and adjoining E. C. Merewether, Esq.'s, Burwood coal property, a distance of about 5 miles from the Newcastle Harbour. It is 15 feet 7 inches in thickness, 7 feet of which it is proposed to work. It is a free-burning bituminous coal, suitable for household, steam, smelting, gas, coking, and blacksmith purposes; has an excellent roof and floor, and dips about one in twenty to the south-west. There are several other seams of coal known to exist here; but the one above described is of such an excellent quality, and can be so cheaply and economically worked, that the present proprietors only intend working it at the commencement, and are at present endeavouring to form a Company to work the coal from under the land. Its proximity to the Newcastle Harbour, the excellent quality of the coal being similar to that worked at the other collieries, will, I have no doubt, ensure for it a fair share of the trade, if it is only properly and efficiently opened out and managed.

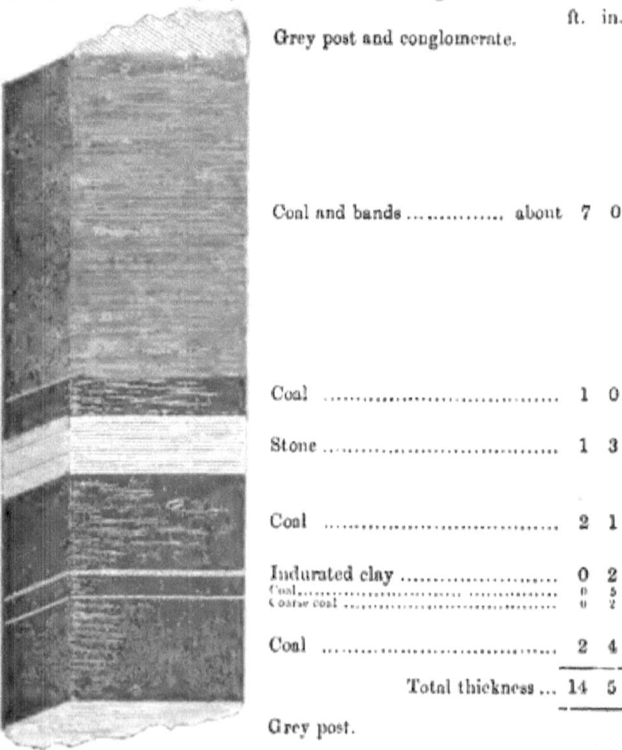

ft. in.
Grey post and conglomerate.
Coal and bands .... about 7 0
Coal 1 0
Stone 1 3
Coal 2 1
Indurated clay 0 2
Coal.................................... 0 5
Coarse coal 0 2
Coal 2 4
Total thickness ... 14 5
Grey post.

This is a section of the No. 2 or Burwood seam of coal worked by Messrs. Gullivar and Ashman, and Messrs. James and Hénry Wilson at the Victoria Tunnel and Glenrock Collieries, the property of E. C. Merewether, Esq. (See L on plan No. 1 and No. 2 on section No. 7.) This seam is 14 feet 5 inches in thickness, 7 feet 5 inches and sometimes only 5 feet 2 inches, is worked by adits driven into the hill side. It lies very flat, with a southerly dip, and the coal is sold for household and other purposes at Burwood and Newcastle. It is a free-burning bituminous coal, suitable for household, steam, smelting, gas, coking, and blacksmith purposes, and the specific gravity of it is about 1·25 to 1·3.

The quantity of coal sold from these two land sale collieries in 1874 was 3,548 tons, valued at £1,862 14s., aud the average number of men employed when the mine was working was sixteen.

Mr. Merewether has lately leased 1,400 acres of this valuable property to J. B. Winship, Esq., and others, who are now making the necessary preparations for opening out and working on a large scale the No. 6 or Borehole seam of coal. It has been proved by boring in several places, and is now being worked by the Australian Agricultural Company to the north and at no very great distance from the northern boundary of the portion leased.

The property is situated only about 2½ miles from the Government cranes and the Newcastle Harbour; the coal is proved to be of excellent quality on the north and east of the portion leased, therefore there is no doubt in my mind that under Mr. Winship's management, and as it is the second nearest colliery to the cranes and harbour, it will be a successful undertaking.

C. J. Stevens, Esq., M.L.A., with his usual enterprise, is now putting down tubbing and sinking to the Borehole seam of coal at Stockton, on the North Shore, immediately adjoining the Newcastle Harbour, where the coal has been proved by borings several years since.

The only difficulty there will be to contend with in working the coal will be the water which is almost sure to be met with in considerable quantities as the mine is extended and opened out, and it will require exploring headings to be kept some distance in advance of the men's working places. Next year I hope I may be able to report that the Borehole seam of coal has been sunk through.

Murrurundi Petroleum Oil Cannel Coal Land.

(Prospected previous to 1874.)

A seam of petroleum oil cannel coal was, previous to 1874, opened out at Colley Creek, about 130 miles north-west from Newcastle, and a few miles distant from the Northern Railway now being constructed past Murrurundi.

Where opened out and examined by me in November, 1872, it was 1 ft. 2 in. of rich petroleum oil cannel coal, equal to the best Hartley, but with a very heavy dip.

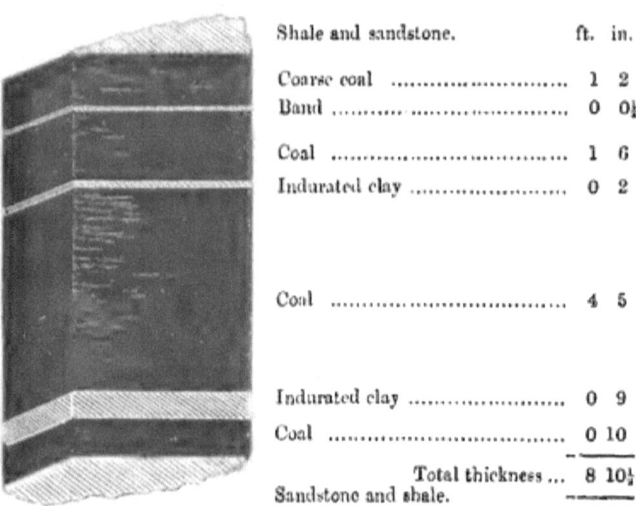

Shale and sandstone.	ft.	in.
Coarse coal	1	2
Band	0	0½
Coal	1	6
Indurated clay	0	2
Coal	4	5
Indurated clay	0	9
Coal	0	10
Total thickness ...	8	10½
Sandstone and shale.		

This represents the 8 feet 10 inches seam of coal now being worked and reopened out by Messrs. O'Brien and Curlewis, at the Woodford Colliery. (See M on plan No. 1.) A short tramway connects the mine with the Great Northern Railway, at a distance of 14 miles from the Newcastle Harbour. There are several shafts sunk on the property, and two seams of coal are worked, one 5 feet 6 inches in thickness, and a lower seam, the one above described, 8 feet 10 inches thick. It is a splint and bituminous coal, suitable for household, steam, smelting, gas, and blacksmith purposes; the samples I have tested giving an average specific gravity of 1·3. The coal and strata dip about 1 in 25 to the south-east.

Messrs. O'Brien and Curlewis are the proprietors, and manage the colliery themselves.

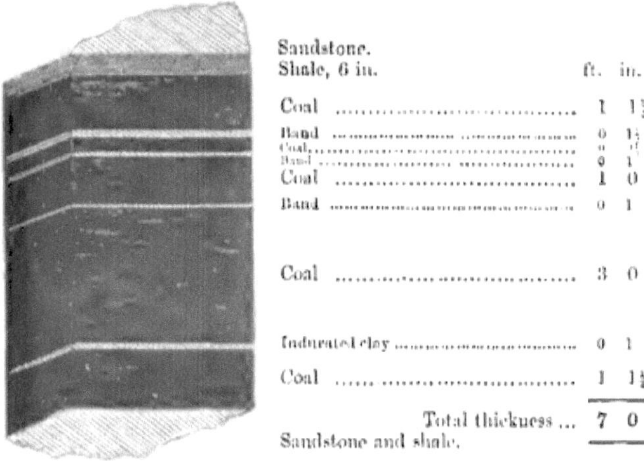

Sandstone.
Shale, 6 in.
ft. in.
Coal ... 1
Band ... 0
Band ... 0 1
Coal ... 1 0
Band ... 0 1
Coal ... 3 0
Indurated clay ... 0 1
Coal ... 1 1½
Total thickness ... 7 0
Sandstone and shale.

This is a section of the 7 to 8 feet seam of coal worked at the Four-mile Creek Collieries (see N on plan No. 1), near Maitland, about 7 miles from Morpeth, the head of the navigation of the Hunter River. It is a splint and bituminous coal, specific gravity about 1·3, suitable for household, steam, smelting, gas, coking, and blacksmith purposes; the mines at work in this district being Messrs. Pearse & Co., Iugaree, Sunderland, Bloomfield, and Dark Creek. The coal is sold to the Hunter River Steam Navigation Companies at Morpeth, and for household and other purposes in the Maitland and Morpeth districts.

The quantity of coal sent away from these collieries in 1874 was 19,053 tons, valued at £6,395 19s.; and twenty-five men and seven boys daily employed when at work.

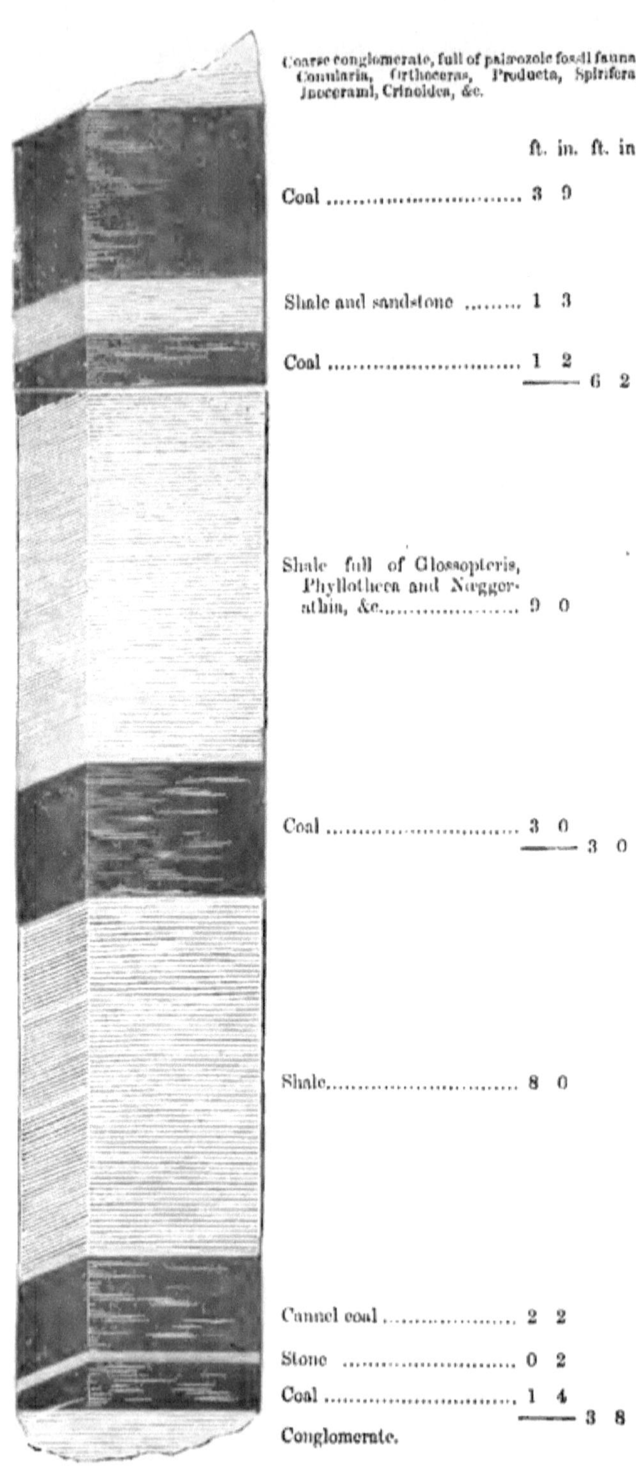

Coarse conglomerate, full of palæozoic fossil fauna, Conularia, Orthoceras, Producta, Spirifera, Inocerami, Crinoidea, &c.
ft. in. ft. in.
Coal 3 9
Shale and sandstone 1 3
Coal 1 2
6 2
Shale full of Glossopteris, Phyllotheca and Næggerathia, &c. 9 0
Coal 3 0
3 0
Shale.............................. 8 0
Cannel coal 2 2
Stone 0 2
Coal 1 4
3 8
Conglomerate.

This represents a section of the seams of coal formerly worked by the Honorable Bourne Russell, at Stony Creek (see O on plan No. 1), and now by Mr. Harpur, for household purposes, in Maitland and the surrounding districts.

Here we have splint, bituminous, and cannel coals, with lower carboniferous (palæozoic) fossil fauna in the shales and rocks above and below these seams of coal, which will be hereinafter referred to. The coal and strata dip at the rate of 1 in 2½ to the south-east, which makes these seams of coal more difficult and expensive to work than those before described.

The coal is of very good quality, and suitable for all purposes.

The quantity sold in 1874 was 500 tons, valued at £300, and the number of men employed, 7.

Conglomerate and sandstone, full of palæozoic fossil fauna, Conularia, Orthoceras, Producta, Spirifers, Inocerami, Crinoidea, &c.

	ft.	in.
Coal	2	6
Indurated clay	0	3½
Coal	3	3
Indurated clay	0	2
Coal	1	6
Stone band	0	3
Coal	0	6
Stone band	0	1
Coal	6	0
Total thickness worked	14	6½

Here we have a section of the seam of coal worked at the Anvil Creek Colliery. (See P on plan No. 1), adjoining the Great Northern Railway, and connected therewith by a short tramway at a distance of 32 miles from the Newcastle Harbour.

It is 14 feet $6\frac{1}{2}$ inches in thickness, dips at the rate of 1 in 9 to the west, and is worked from a shaft 100 feet in depth.

It is a splint coal of excellent quality, suitable for steam, household, smelting, gas, and other purposes, specific gravity about 1·25, and from its hard splinty nature, it is a coal that bears transhipment and carriage well, as it will not readily break and form itself into slack.

This Company was formed last year, and their yield of coal is steadily on the increase, and I understand it meets with a ready sale in foreign and intercolonial markets. They have one 10 h-p. winding engine, and seventy coal waggons for taking the coal from the mine to the Newcastle Harbour.

The quantity of coal raised in 1874 was 21,000 tons, valued at £16,800, and they employed daily when working, ninety miners, thirteen labourers and others, and ten boys.

Mr. Swinburn is now the Colliery Manager, and J. B. Winship, Esq., the Colliery Viewer.

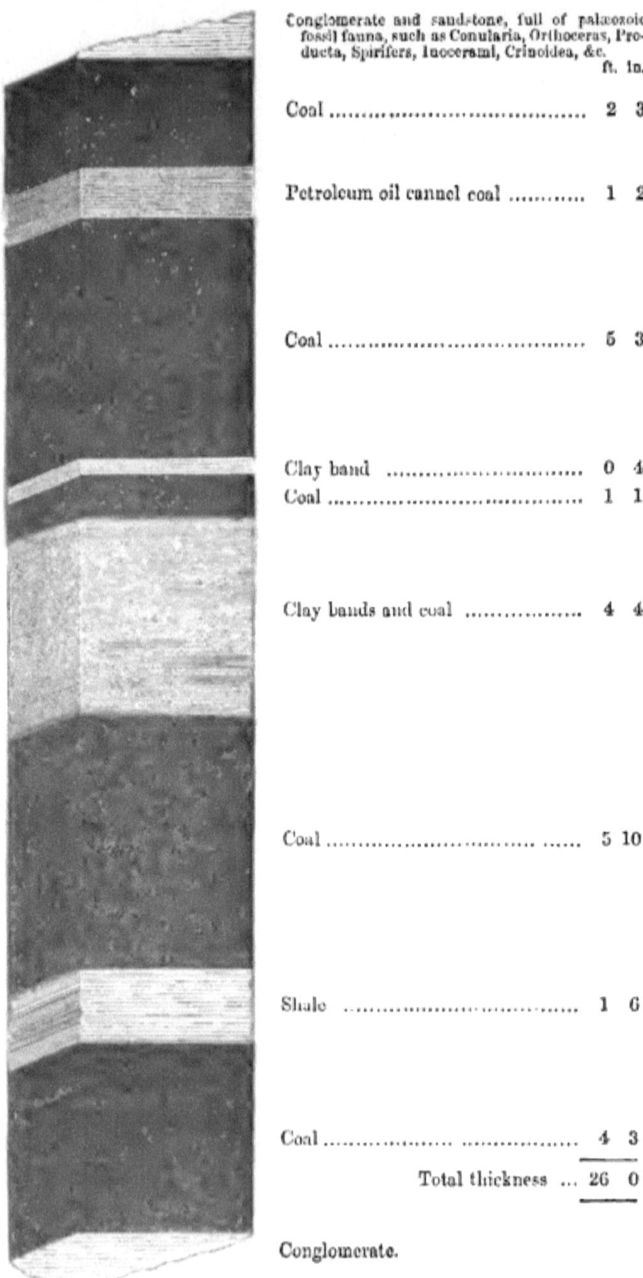
Conglomerate and sandstone, full of palæozoic fossil fauna, such as Conularia, Orthoceras, Producta, Spirifers, Inocerami, Crinoidea, &c.
ft. in.
Coal 2 3
Petroleum oil cannel coal 1 2
Coal 5 3
Clay band 0 4
Coal 1 1
Clay bands and coal 4 4
Coal 5 10
Shale 1 6
Coal 4 3
Total thickness ... 26 0
Conglomerate.

This is a section of the seam of coal worked at the Greta Coal and Shale Company's Colliery adjoining the Great Northern Railway situated 32 miles by rail from the Newcastle Harbour (see Q on plan No. 1), where it is 26 feet in thickness. The Company hold 2,136 acres of land, and where this seam of coal was first opened out, and where they are now working it from a shaft 108 feet in depth, there is 1 foot 2 inches of very rich petroleum oil cannel coal, between coal, near the top of the seam, but it has proved to be only of a limited area. It is a splint coal, of excellent quality, suitable for steam, household, smelting, gas, and other purposes, specific gravity about 1·25, and from its hard splinty nature it is a coal that bears transhipment and carriage well, as it does not readily break and form itself into slack. The Company are extending their operations by sinking a second shaft nearly half a mile on the deep of their No. 1 shaft where they lately won the coal 17½ feet in thickness at a depth of 450 feet, and found the usual palæozoic fossil fauna over the coal in the rocks sunk through, which fossils are similar to those so often referred to by the Rev. W. B. Clarke, M.A., F.G.S. &c., &c., myself and others, as being found between our upper and middle, and middle and lower measures. The coal has a strong rock roof and floor, and dips at the rate of about 1 in 6 to the west. Alfred Vindin, Esq., is the Manager, and James Fletcher, Esq., the Colliery Viewer.

Edwd. Campbell, Esq's., Coal Mine, at Rix's Creek, Singleton. This seam of coal has not been opened out in a sufficient number of places to enable me to give an accurate average section of it; where opened it is about 16 feet in thickness with a great number of bands, and the portion worked has generally been about 4 feet. It is a splint and bituminous coal, suitable for household, steam, smelting, gas, and blacksmith purposes, and is sold in the neighbourhood of Singleton. The quantity of coal sold in 1874 was 180 tons, valued at £126, and two men and one boy were employed.

Southern District.

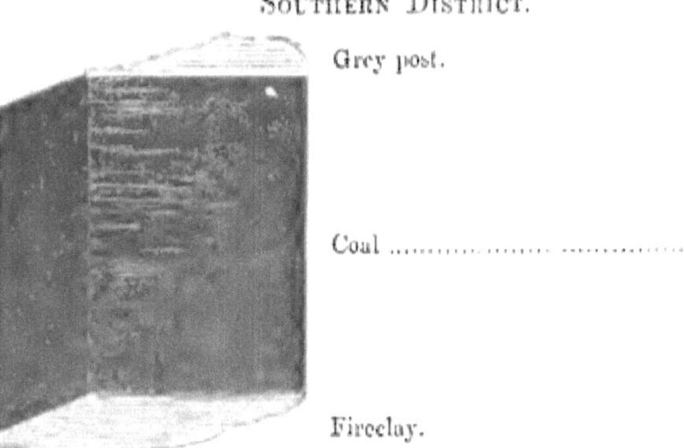

ft. in.

Grey post.

Coal 5 6

Fireclay.

This denotes a section of the seam of coal 5 feet 6 inches in thickness (See section lettered J, section No. 16), formerly belonging to Campbell Mitchell, Esq., and now to His Honor Mr. Justice Hargraves, at Coal Cliff, in the County of Cumberland, and adjoins the extensive selections of Messrs. Alexander Stuart and Company. It is a semi-bituminous coal, specific gravity about 1·3, has a slight dip to the north-east, an excellent roof and floor, would bear carriage well, and is a good steam, smelting, and household coal. It lies about 40 feet above the sea level, and in close proximity to a proposed dock and the projected South Coast Railway, which has been surveyed through the property and will pass close by the entrance to the mine.

Grey post.

Coal .... ft. 8

Fireclay.

This represents a section of the seam of coal worked at the Bulli Colliery, near Wollongong, in the county of Camden, situated about 40 miles south of Port Jackson. It averages about 8 feet in thickness of clean coal, without any bands in it; the average specific gravity is about 1·3; it has an excellent rock roof and floor; and the dip is about 1 in 30 to the north-west.

It is a semi-bituminous coal, suitable for steam, household, smelting, blacksmith, and other purposes, and is by many captains of vessels used in preference to the more bituminous coal of the Newcastle District, and being a stronger coal the firebars have to be put further apart when using it.

This is the No. 1 or uppermost seam of coal in the district (see letter H on plan No. 5 and section No. 14), and is found outcropping

PLAN No. 2

Referred to, showing the position and extent of the Bulli Colliery Co.'s Mine, situated 40 miles south of Port Jackson, in the

COUNTY OF CAMDEN.

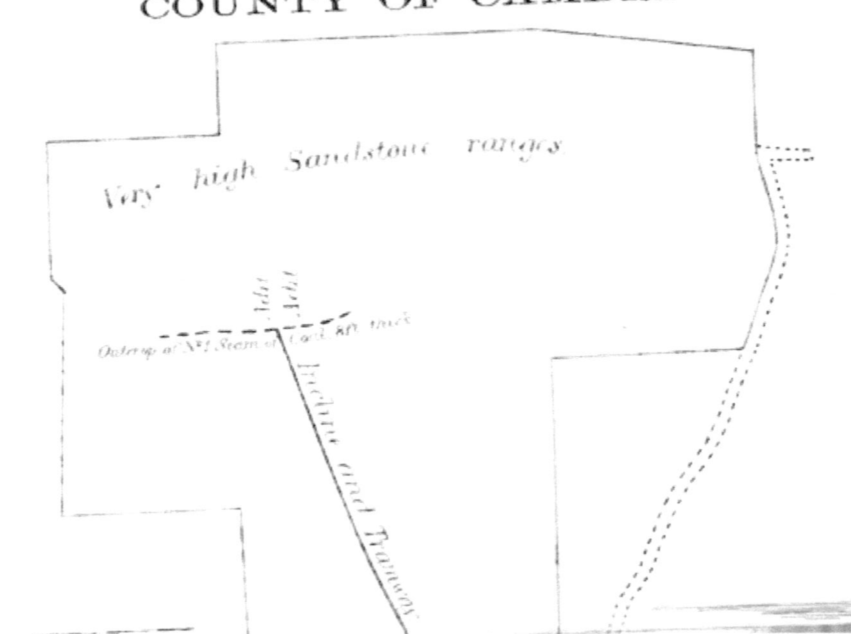

OCEAN
Bulli Colliery Jetty

at a height of about 400 feet above the sea in the high ranges fronting the Pacific Ocean, where adits are driven into it, and an incline and tramway, 2¼ miles in length, conveys the coal from the mine to the Company's jetty, where steam colliers come alongside and receive the coal. (See plan No. 2). This Company sent away 58,506 tons of coal in 1874, and employed daily, when at work, eighty-five miners and sixty-five labourers and others. Mr. Ross is the Colliery Manager, J. B. Winship, Esq., the Colliery Viewer, and Jas. Shoobert, Esq., the General Manager and Superintendent.

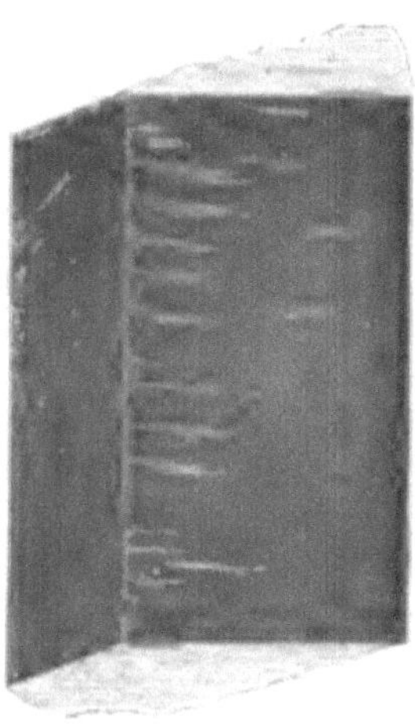

Grey post.

	ft.	in.
Coal	7	6

Fireclay.

This is a section of the seam of coal worked at the Mount Pleasant Colliery, near Wollongong, in the county of Camden, situated 50 miles to the south of Port Jackson. It averages about 7 feet 6 inches in thickness of clean coal, without any bands in it; the average specific gravity is about 1·3; it has an excellent rock roof and floor; and the dip is about 1 in 30 to the west and north-west.

It is a semi-bituminous coal, suitable for steam, household, smelting, blacksmith, and other purposes, and is by many captains of vessels used in preference to the more bituminous coal of the Newcastle District, and being a stronger coal the firebars have to be put further apart when using it. This is the No. 1 or uppermost seam of coal in the district (see letter H on plan No. 5 and section No. 14), and is found outcropping at a height of about 600 feet above the sea level in the high ranges fronting the Wollongong Harbour, where adits are driven into it, and an incline and tramway, about 3 miles in length, takes the coal from the mine to the Wollongong Harbour. (See plan No. 3). The harbour is about 440 feet in length and 150 feet in breadth, with an average depth at low-water of 14 feet; under the coal staiths it is 18 feet, so that vessels drawing that depth can enter at high-water and load under the staiths without hindrance.

The quantity of coal sent away from this colliery in 1874 was 38,985 tons, valued at £16,568; and when the colliery was at work they employed an average of 120 men and boys, and had twenty-eight horses, most of the latter being employed in taking the coal from the bottom of the incline to the harbour. Patrick Lahiff, Esq., is the Colliery Manager.

Grey post.

	ft. in.
Coal	7 6

Fireclay.

Here we have a section of the seam of coal worked at the Osborne Wallsend Colliery, near Wollongong, in the county of Camden, belonging to J. Osborne, Esq., and situated about 50 miles south of Port Jackson. (See plan No. 3.) It averages about 7 feet 6 inches in thickness of clean coal, without any bands in it; the average specific gravity is about 1·3; it has an excellent rock roof and floor; and the dip is about 1 in 30 to the west and north-west.

It is a semi-bituminous coal, suitable for steam, household, smelting, blacksmith, and other purposes, and is by many captains of vessels used in preference to the more bituminous coal of the Newcastle district, and being a stronger coal the firebars have to be put further apart. This is the No. 1 or uppermost seam of coal in the district (see letter H on plan and section No. 14), and it is found outcropping at a height of about 600 feet above the sea-level, in the high ranges fronting the Wollongong Harbour, where

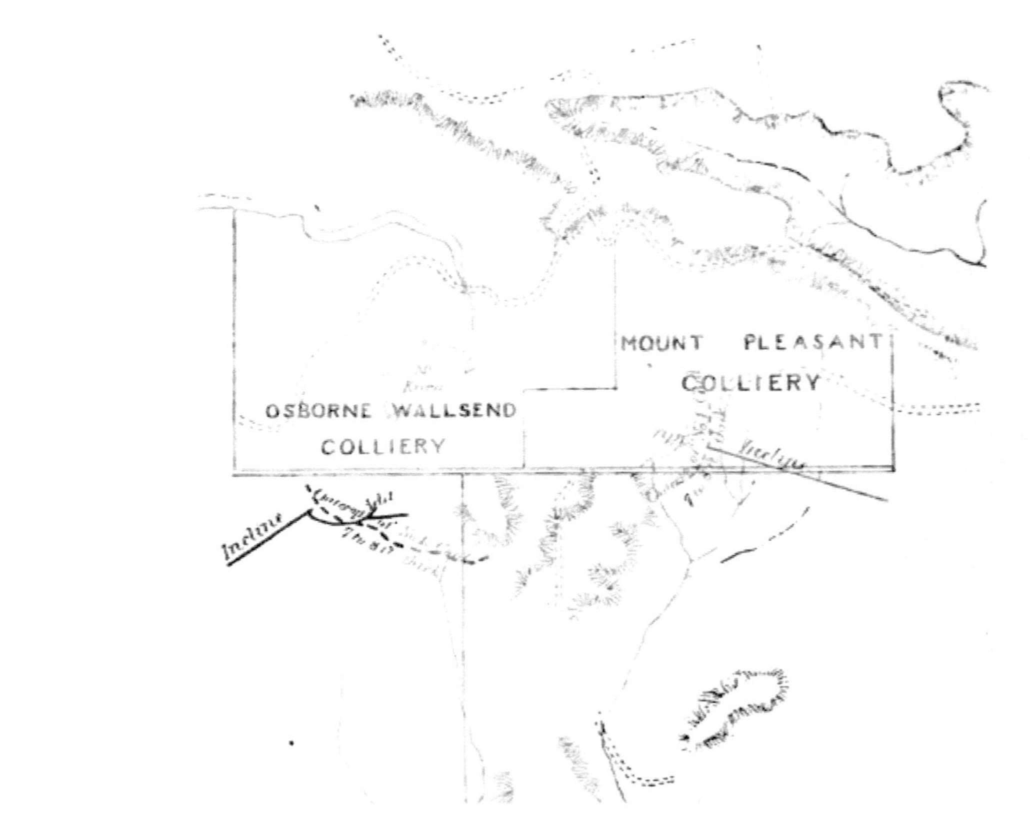

MOUNT PLEASANT
COLLIERY
OSBORNE WALLSEND
COLLIERY
Incline
Incline

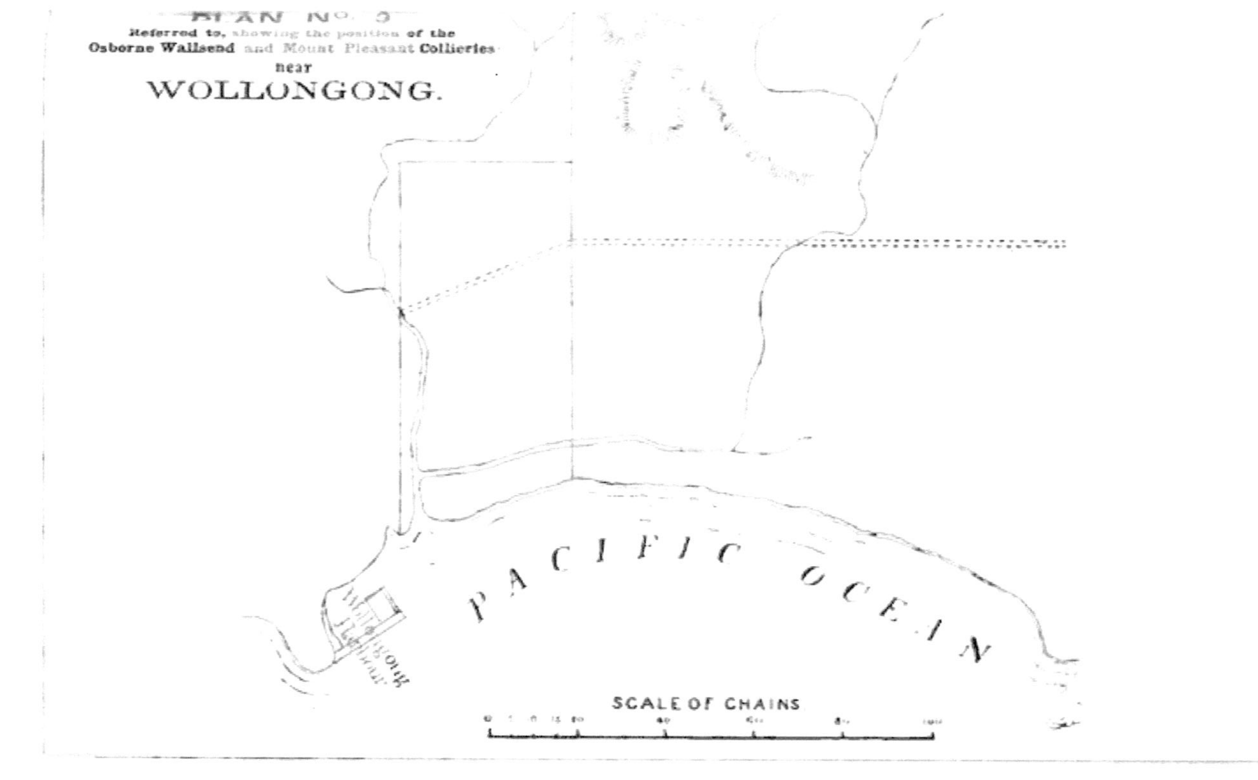
PLAN No. 3
Referred to, showing the position of the
Osborne Wallsend and Mount Pleasant Collieries
near
WOLLONGONG.
PACIFIC OCEAN
Wollongong Harbour
SCALE OF CHAINS
0 5 10 15 20 40 60 80 100

adits are driven into it, and an incline and horse tramway, of about 2½ miles in length, conveys the coal from the mine to the Wollongong Harbour before described. The quantity of coal sold from this colliery in 1874 was 37,796 tons, valued at £16,063, and the average daily number of miners, labourers, and others, was ninety-seven, and thirty boys. Mr. Johnson is the Mining Manager, and J. B. Winship, Esq., the Colliery Viewer.

The Mount Kembla Coal and Shale Mines at American Creek, near Wollongong.

Here is a seam, 18 inches to 2 feet in thickness, of petroleum oil schist, which has for many years been worked, and kerosene oil, benzine and lubricating oil, &c., manufactured therefrom by works erected at the mine, and the oil sold in Wollongong, Sydney, and other places.

The quantity of petroleum oil schist used for the manufacture of oil in 1874 was 1,000 tons, valued at £500, and they employed daily on an average fourteen miners, four wheelers, and one overman.

Coal lands in the Southern District which have been prospected in the year 1874.

County of Camden.

Brereton's Coal Mine, near Berrima, has raised 1,000 tons of coal in 1874, valued at £1,250; but no notice of opening out having been sent to this office, I had no idea that it was at work or I should have examined it, and now given a section of the seam, which is, I understand, very thick, and of a good bituminous quality.

Campbell Mitchell, Esq., and others, have taken up and done some prospecting on large areas of land in the neighbourhood of the Fitzroy Iron Mines, near Nattai, on ths Great Southern Railway. Mr. Mitchell informs me that one seam which he has opened out is 12 ft. in thickness, exclusive of bands.

Joadja Creek, in the county of Camden.

Messrs. Brown & Lamb, Larkins, and others, have taken up a considerable quantity of land at Joadja Creek, situated about 12 miles from the Great Southern Railway, near Berrima, where there is a very rich deposit of petroleum oil cannel coal. Mr. Brown kindly accompanied me on my inspection, and pointed out the different provings and drives they had made in the seam.

The following is a section taken in one of the drives, near the centre and thickest part of the deposit :—

	ft.	in.
Sandstone and conglomerate (roof).		
Coal, very bright	0	8
Petroleum oil cannel coal (rich)	1	0
Indurated clay	0	1
Petroleum oil cannel coal (very rich), equal to the best Hartley	1	3
Coal, very bright	1	0
Black shale, 2 inches.		
Rock (floor).		
	2	4

Two measurements I took nearer the edge of the deposit :—

No. 1.

	ft.	in.
Sandstone (roof).		
Coal	0	4
Petroleum oil cannel coal, very rich, equal to the best Hartley	1	8
Coal	0	11
Sandstone (floor).		
Total thickness...	2	11

No. 2.

	ft.	in.
Sandstone (roof)	50	0
Petroleum oil cannel coal, very rich, equal to the best Hartley...... Total thickness...	1	6
Sandstone (floor).		

On Mr. Larkins' land, where he had opened it out, it measures :—

	ft.	in.
Sandstone and conglomerate (roof).		
Inferior bituminous shale, full of Glossopteris	0	2
Petroleum oil cannel coal	1	5
Inferior bituminous shale	0	3
Sandstone (floor).		
	1	10, or

1 ft. 5 in. of petroleum oil cannel coal. Thirty yards from where this measurement is taken the cannel has decreased to 10 inches, with a rock roof and floor. The dip is about 3½° north 20 west.

Jamberoo, near Kiama, in the county of Camden.

Large areas of coal land have been selected by Messrs. Parkes, Sutherland, Nichols, Lord, and others, near Jamberoo, and the Saddle-back Ranges, a few miles from the Kiama Harbour, and about 20 miles from Jarvis Bay, the largest sheet of sheltered waters in New South Wales.

Messrs. Parkes, Sutherland, and Nichols, have opened out two thick and excellent seams of coal in several places on their land, and that adjoining which I examined and reported upon for them in January last. The minerals proved to exist here are coal, hematite, clay band, and brown hydrated oxide of iron ores, and limestone. The uppermost or No. 1 seam of coal near the north-eastern portion of these selections is 39 feet in thickness, and 10 feet 6 inches of the lower part is what I consider to be the best and workable portion of it.

The second is about 24 feet in thickness, the lower 10 feet appearing as if it would be the best and workable portion, but before this can be accurately ascertained it requires to be driven into, which had not been done when I made my examination. The seams of coal vary very much in their thickness and character in this neighbourhood.

The iron ores above mentioned exist here in large quantities, and some of the specimens of brown hydrated iron ore yield $51\frac{1}{2}$ per cent. of metallic iron. The limestone lies in a thick bed below the thick seams of coal before mentioned.

I have seen no workable seams of coal south of Jarvis Bay, and there have been none found up to the present time south of these selections. These belong to the upper coal measures, and have the usual palæozoic or lower marine Carboniferous fauna underneath in strata lying conformable with them.

Port Hacking, in the county of Cumberland.

1,946 acres of land has been leased here and formed into a Melbourne Company, with a capital of £100,000, for the purpose of boring for coal, and with the expectation of finding it at a depth of 200 to 300 feet at Port Hacking, which is about 15 miles south of Port Jackson. Two of the Directors called on me a short time since to know if I would tell them at what depth I thought they would find the first seam of coal likely to be of a workable thickness. I replied that instead of 200 to 300 feet, it would be most probably be more than 1,000 feet, and in some places on the land selected possibly about 1,500 feet, and that if they intended to bore they should start with a good-sized borehole, so as to enable them to go down at least 1,000 feet.

Western District.

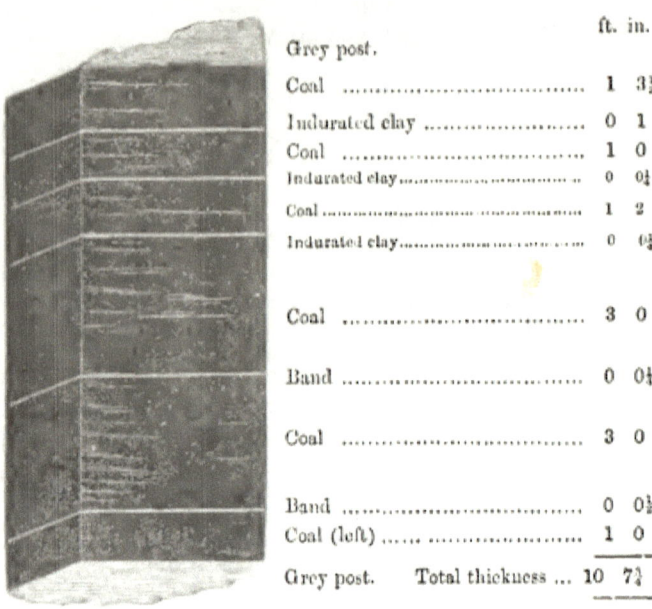

	ft.	in.
Grey post.		
Coal	1	3½
Indurated clay	0	1
Coal	1	0
Indurated clay	0	0¼
Coal	1	2
Indurated clay	0	0¾
Coal	3	0
Band	0	0¼
Coal	3	0
Band	0	0½
Coal (left)	1	0
Grey post. Total thickness ...	10	7¾

This is a section of the seam of coal lately sunk to at the Vale of Clwydd Colliery, in Lithgow Valley, in the county of Cook. (See R on plan No. 4, and section D No. 10.) The shaft is 233 in depth, and the coal is 10 feet 7¾ inches in thickness, with a dip to the east of about 1 in 20. This Company have 323 acres, and the shaft is connected, by a tramway of less than a quarter of a mile in length, with the Great Western Railway, at a distance of 95 miles from Sydney and the Harbour of Port Jackson. It is a splint coal, well suited for steam, household, smelting, gas, blacksmith, and other purposes; specific gravity about 1·3, and from its hard splinty nature it is a coal that will bear carriage well, as it will not readily break and form itself into small coal. The quantity of coal got in December, 1874, was 50 tons, valued at £17 10s., and the average number of men employed daily was about sixteen and one boy. Mr. Wilson is the Colliery Manager, and A. Fairfax, Esq., the General Manager.

PLAN N°. 4

Referred to, showing the position and extent of the different Collieries now at work at Lithgow Valley, in the

COUNTY OF COOK.

WESTERN DISTRICT.

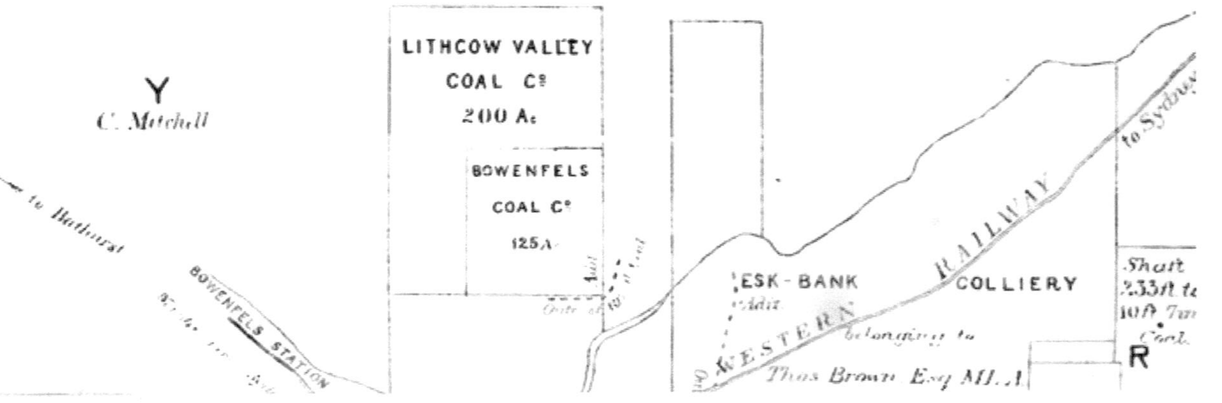

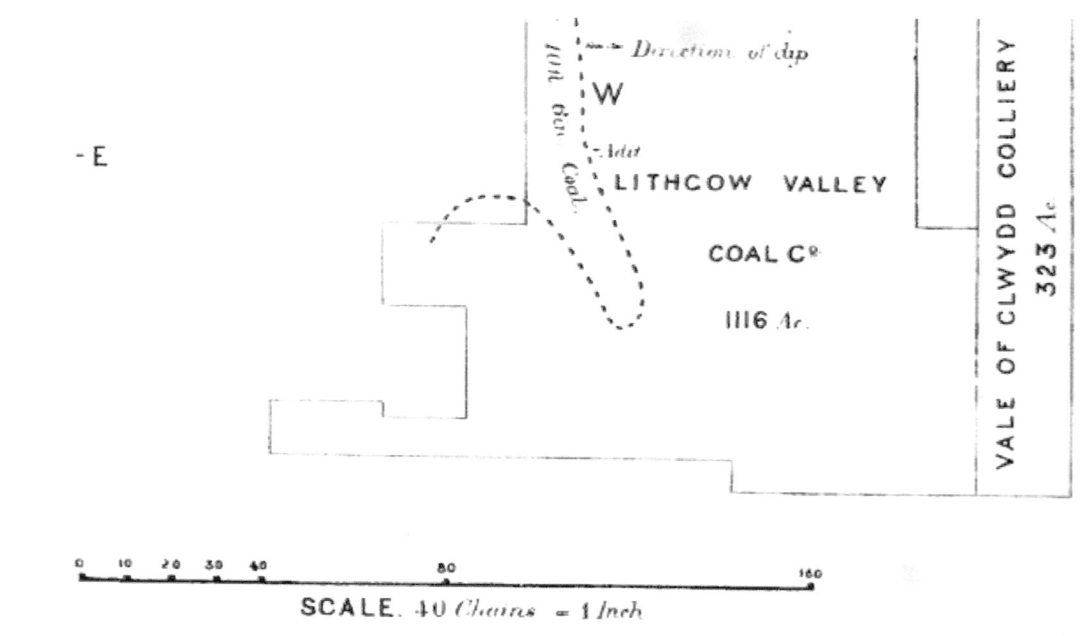

SCALE. 40 Chains = 1 Inch

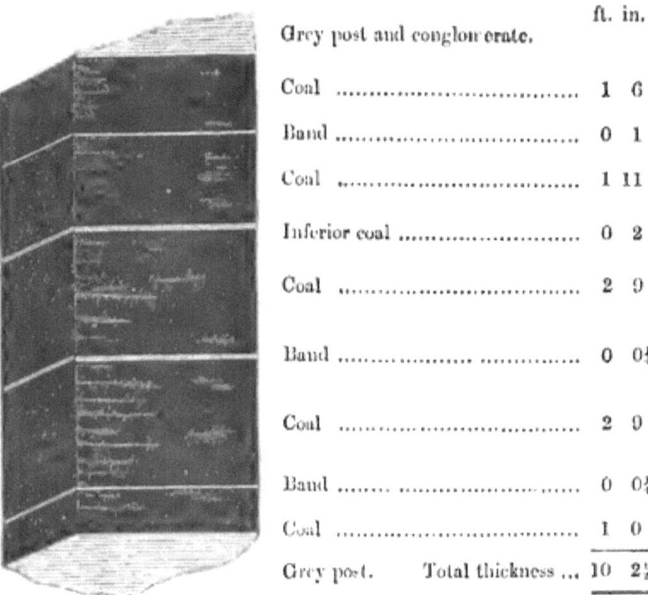

	ft.	in.
Grey post and conglomerate.		
Coal	1	6
Band	0	1
Coal	1	11
Inferior coal	0	2
Coal	2	9
Band	0	0¼
Coal	2	9
Band	0	0¼
Coal	1	0
Grey post. Total thickness ...	10	2½

This is the seam of coal worked by Thomas Brown, Esq., M.L.A., at the Eskbank Colliery, in Lithgow Valley, in the county of Cook. (See 8 on plan No. 4, and lower coal on section lettered D No. 10.) The property consists of 740 acres of freehold land, and the coal is worked from a shaft 83 feet 6 inches in depth, situated close to and alongside the Great Western Railway, at a distance of about 96 miles from Sydney. It is a splint coal, which has been found well suited for steam, household, smelting, gas, blacksmith, and other purposes, dips about 1 in 20 to the east, and from its hard splinty nature has been found to bear carriage well, as it does not readily break and form itself into slack. The quantity of coal sold at this colliery in 1874 was 8,600 tons, valued at £3,010, and the average daily number of men employed when the pit was working was fourteen.

Furnaces for smelting copper ores from the western district are erected on this property, and an iron smelting furnace is also now in course of construction.

The Eskbank Colliery Copper Smelting Works are managed by Mr. Lloyd, who erected the Cow Flat Copper Smelting Works. The Company have 15½ acres of land leased from Mr. Brown. These works consist of three smelting and one refining furnace, and are erected about 70 yards from the 83 feet 6 inch coalpit before referred to on this property. The furnaces were only working continuously for the last three months of 1874. The following is the quantity of regulus from the Cow Flat Copper Mine, and copper ores from Wiseman's Creek, near Bathurst, and a mine near Goulburn, which have been smelted during 1874:—

	tons
Regulus from Cow Flat Copper Mining Company, about	420
Copper ore from Wiseman's Creek, near Bathurst	160
Do. do. near Goulburn	15

The Manager informs me that the output of refined copper for the last three months, ending February, was 181 tons, or 14 tons per week, the whole of which has been forwarded to Sydney for shipment to England, in parcels of 20 to 30 tons. The works have been erected with bricks made from clay on the Company's land, but there is far better fireclay to be found on this property and adjoining the part leased to this Company. Mr. Brown has contracted to deliver coal to the furnaces for fifty years at 2s. 6d. per ton for small coal and 6s. for round.

Iron Smelting.

One iron smelting furnace is now in course of construction on 100 acres of land leased from Mr. Brown, by Messrs. Hughes, Sutherland, Williams, Rutherford, Denny, and Kelly, and fire-brick works have also been erected. The smelting furnace is being erected on the rock floor of the thick seam of coal, and is in close proximity to Farmer's Creek, where there is a never-failing supply of water. The foundation, and a portion of the brickwork, was in February last completed for a furnace 50 feet in height and 12 feet across the boshes, which Mr. Hughes calculates will produce 120 tons of iron per week, and that they will be making iron about next September. There are numerous bands of argillaceous iron ore on this property and that of the other collieries adjoining, similar to that found and worked in England, and Mr. Hughes informs me that some of it which he had tested in Bathurst yielded 39 to 49 per cent. of metallic iron. It is proposed to bring magnetic and brown hydrated oxide of iron ores from Mount Lambie, where the Company have 200 acres of land on mineral lease, and mix it with the clay band iron ores. They also intend to bring red hematite from their 60 acres of land at Mount Victoria, and limestone will be brought from Blayney or Piper's Flat, near Wallerawang. Mr. Hughes has erected a brick-kiln capable of burning 25,000 bricks, and fireclay of a very excellent quality is procured from a bed 6 feet in thickness on the land where a Chilian mill has been erected for grinding the clay. The brick-shed is 120 feet by 25 feet, in which 100,000 common building bricks have been made, and a large quantity of very superior firebricks, and lumps and large circular bricks 3 feet in length for lining the furnace. Enoch Hughes, Esq., is the Manager.

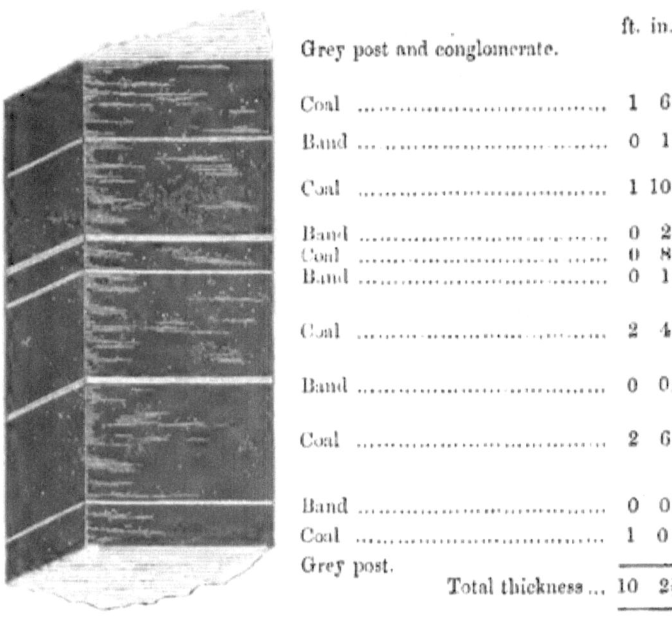

	ft.	in.
Grey post and conglomerate.		
Coal	1	6
Band	0	1
Coal	1	10
Band	0	2
Coal	0	8
Band	0	1
Coal	2	4
Band	0	0⅛
Coal	2	6
Band	0	0⅛
Coal	1	0
Grey post.		
Total thickness ...	10	2¼

Here we have a section of the seam of coal worked at the Lithgow Valley Company's Colliery, situated about 96 miles by rail from Sydney and the harbour of Port Jackson. Their property consists of 1,116 acres. (See letter W on plan No. 4). The coal is here about 10 feet 2¼ inches in thickness, and is worked by an adit driven into the seam, which is connected with the Great Western Railway by a tramway less than ½ a mile in length. It is a splint coal, which has been found well suited for steam, household, smelting, gas, blacksmith, and other purposes, dips about 1 in 20 to the east, and from its hard splinty nature has been found to bear carriage well, as it does not readily break and form itself into slack. When I visited this colliery last they were opening out the upper thick coal for the purpose of making coke. It is a more bituminous coal, but is very full of bands.

The quantity of coal sold from this colliery in 1874 was 18,000 tons, valued at £5,400, and the average number of men daily employed when the colliery was at work was twenty-five. Mr. Douglas is the Colliery Manager.

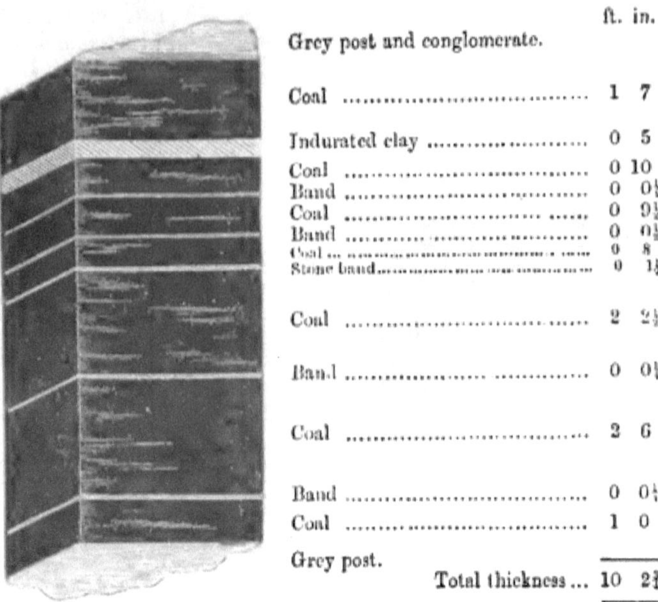

	ft.	in.
Grey post and conglomerate.		
Coal	1	7
Indurated clay	0	5
Coal	0	10
Band	0	0½
Coal	0	9½
Band	0	0½
Coal	0	8
Stone band	0	1½
Coal	2	2½
Band	0	0¼
Coal	2	6
Band	0	0¼
Coal	1	0
Grey post.		
Total thickness ...	10	2¾

This represents the seam of coal 10 feet 3 inches in thickness, worked by the Bowenfels Coal Mining Company at Lithgow Valley, situated about 97 miles by rail from Sydney and the harbour of Port Jackson. Their property consists of 125 acres of mineral land (see X on plan No. 4), and the coal is worked from an adit driven into the hill-side, from which a tramway of about ½ a mile in length takes the coal on to the Great Western Railway, and thence to Sydney and elsewhere. It is a splint coal, which has been found well suited for steam, household, smelting, gas, blacksmith, and other purposes, dips about 1 in 20 to the east, and from its hard splinty nature bears carriage well, as it does not readily break and form itself into slack. It is the intention of this Company to erect smelting works on the land.

The quantity of coal sold from this colliery in 1874 was 8,500 tons, value not given. They employed daily when the colliery was at work ten men. Mr. Hepple is the Colliery Manager.

These four Western coal sections represent one and the same seam (see the lowest on section lettered D No. 10), and have other beds of coal, fireclay, and bands of iron ore, in the strata lying over this coal. When the railway is completed past Bathurst to within a few miles of the rich western copper lodes we shall see these collieries raising a much larger quantity of coal, and no doubt other mines opened out between here and Piper's Flat. The coal is used by the Government on their railway, and if they only reduced the rate of carriage for coal to Sydney and the suburbs, the consumption of coal in that direction would be very largely increased. It is well liked in Sydney for household purposes by many who I hear are in the habit of using it.

	ft.	in.
Grey post and conglomerate.		
Coal	0	8½
Indurated clay	0	1
Coal	3	8
Indurated clay	0	0¼
Coal	0	7
Hard blue shale—3 inches.		
	5	0¾

This is a section of the same seam of coal where it has decreased to 5 feet 0¾ inch in thickness. It is situated on land belonging to Campbell Mitchell, Esq., opposite the Bowenfels Railway Station, at a distance of about 97 miles by rail from from Sydney and the harbour of Port Jackson. It is a splint coal similar to that before described, and is suitable for similar purposes, and the position of it is shown by letter Y on plan No. 4.

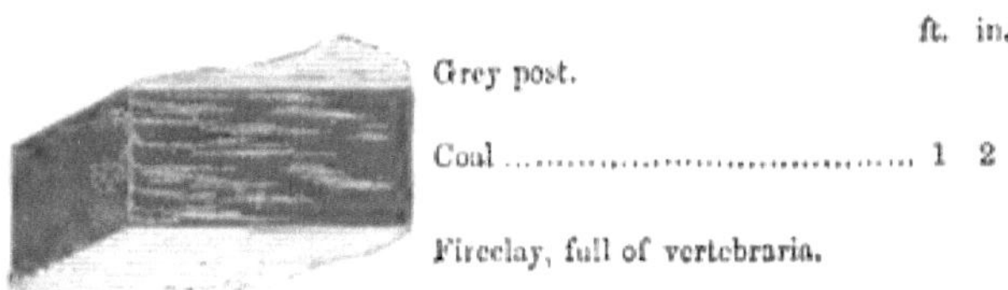

	ft.	in.
Grey post.		
Coal	1	2
Fireclay, full of vertebraria.		

This denotes a section of the same seam of coal at Z, on plan No. 4, where it has decreased from over 10 feet in thickness to 1 foot 2 inches, which is no uncommon occurrence either here, in England, or any other coal country.

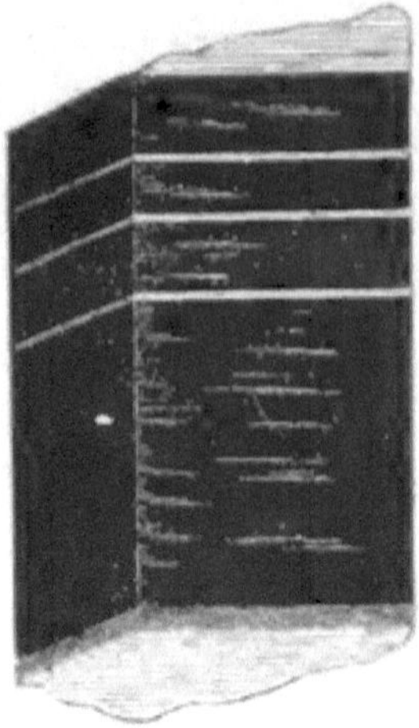

	ft.	in.
Grey post and conglomerate.		
Coal	1	1
Indurated clay	0	1
Coal	0	9
Indurated clay	0	0¼
Coal	1	0
Stone band	0	2
Coal	4	2
Total thickness	7	3¼
Black shale and clay.		

This represents the section of the same seam of coal opened opened out lately by Mr. Bulkeley, at Blackman's Flat, on the Mudgee Road, about 4 miles from the Wallerawang Railway Station, in the county of Cook, and about 12 miles north of Lithgow Valley. It is here 7 feet 3¼ inches in thickness, dips to the west, and is of a similar quality to that I have before described at Lithgow Valley. The coal is carted to Wallerawang Railway Station, and sold in Bathurst. They have 100 acres of land, sold 50 tons in 1874, and employed three men.

New South Wales Shale Oil Company, Sydney, and Hartley in the county of Cook.

Here we have one of the richest seams of petroleum oil cannel coal ever worked here, or its equal found in any other part of the world. It yields 150 to 160 gallons of crude oil to the ton, and 18,000 cubic feet of gas per ton, with an illuminating power equal to 40 candles. This peculiar and exceedingly valuable mineral has a brown streak when scratched, a specific gravity of 1·065, is extremely tough, and when struck with a hammer the instrument rebounds from it somewhat similar to hitting a piece of wood.

The seam measures 5 feet at the thickest part in their mine, situated at Hartley, 80 miles west of Sydney; the centre, 3 feet, being the richest, and that which yields the 18,000 cubic feet of gas to the ton. Thus—

	ft.	in.
Bituminous shale (tops)	1	0
Petroleum oil cannel coal, yielding 160 gallons of crude oil to the ton	3	0
Petroleum oil cannel coal (bottoms)	1	0
	5	0

Large quantities of the mineral are sent to different gas works here, the adjacent Colonies, China, San Francisco, &c., to mix with common coal for the purpose of increasing the illuminating power of the gas.

The Company have very large works erected at Waterloo, near Sydney, where they manufacture kerosene, lubricating and other oils from the mineral which is brought by the Great Western Railway from the mine, a distance of about 80 miles.

The quantity of petroleum oil cannel coal sent away from the mine in 1874 was 9,000 tons, valued at £22,500, and they employed at their colliery daily when working twenty-nine miners, seven labourers, four boys, and one overman. The Company's offices are in Sydney, and John MacDonald, junr., Esq., is the Secretary.

Coal Lands in the Western District, which have been prospected in the year 1874. County of Cook.

Messrs. Winter and others have during the year developed several large and very rich deposits of magnetic, brown hematite, and clay band iron ores, situated at Piper's Flat, about 6 miles from the Wallerawang Railway Station, which is 105 miles by rail from the Sydney Harbour. They have also opened out thick seams of coal in the neighbourhood of the iron ores, and an excellent bed of limestone.

This property has been reported upon by Professor Liversidge, of the Sydney University, and J. W. Winship, Esq., Colliery Viewer, Newcastle, and it is the intention of the present proprietors to form it into a Company.

The variety and richness of the iron ores, and their proximity to thick and good seams of coal and limestone, makes it a very valuable property.

Campbell Mitchell, Esq., and Mr. Croaker and others, have also taken up large areas of land at Piper's Flat, and opened out seams of coal by adits driven into the hill side, one of which is 4 feet 6 inches in thickness, has a specific gravity of 1·2, and is suitable for household, steam, smelting, gas, coking, and blacksmith purposes.

Dr. Mackenzie and C. K. Moore, Esq., had two men prospecting in 1874, and driving adits into their coal and petroleum oil cannel coal seams at Bathgate, near Wallerawang. This mineral yields 17,000 cubic feet of gas to the ton, and has an illuminating power of not less than thirty-six candles, according to A. Norman Tate, F.G.S., analytical chemist, Liverpool, England.

They also had two men for several months in 1874, prospecting and driving in the petroleum oil cannel coal 18 inches in thickness, belonging to them at the Sugar Loaf, at Mount Victoria, in the county of Cook. The drives are on the edge of the deposit. The centre, being covered by high sandstone ranges, cannot be proved without driving the adits a long distance.

Coal Lands in the Western District, prospected previous to 1874.

The Honble. John Lucas and Sir Jas. Martin and others have a considerable area of coal land at Pulpit Hill near the Great Western Railway, at a distance of about 65 miles from Sydney, where very thick and good seams of coal have been opened out; and at Megalow, near Pulpit Hill, Messrs. Mitchell, Mort, and Brown have land with very rich petroleum oil cannel upon it.

Plan (No. 5) of New South Wales Coal Field and Sections.

I annex a plan (No. 5) of the New South Wales Coal Fields, as far as I know and have traced the boundary.

The area shown thereon is about 15,419 square miles, which I consider once formed one large coal basin, and that since its formation upheavals and disturbances of the strata, near the edge of the basin has there thrown it into a series of anticlinal and synclinal curves. See sketch section (No. 6) drawn for the purpose of illustration, and also to show the position of different fossil flora and fauna found in our upper, middle, and lower coal measures.

Sections No. 7 to No. 16 are ten vertical sections I have taken of the upper coal measures, and the places where the measurements were taken are shown on the plan (No. 5) above referred to by letters similar to those on the top of the sections.

Without any exaggeration, we can undoubtedly claim to be in possession of the richest, most accessible, and most extensive Coal Fields in the Southern Hemisphere, which must ultimately make New South Wales the greatest and richest of all the Australian Colonies; and we know the value of them, and how much as a nation Great Britain has to depend upon her collieries for her great and national prosperity. Our bituminous, semi-bituminous, splint, anthracite, and cannel coals are equal in thickness and in quality to any found in other parts of the World, and we have numerous deposits of petroleum oil cannel coal, some of them superior to any yet found elsewhere. During the last few years the growth of our coal trade has most satisfactorily

SKETCH SECTION *about 30 miles in length from* NEWCASTLE *to past* BC
position of the UPPER, MIDDLE, *and* LOWER COAL MEASURES,
connection th

Upper Newcastle Coal Measures, with Glossopteris, Phyllotheca, Vertebraria, Clasteria, Sphenophyllum, Cyclopteris, Anarthrocanna, with Coniferous trees, seed-vessels of Conifers, and fluted stems very like Calamites, and numerous other plants not yet determined by any Palæontologist, and a Urosthenes Australis, (fish).

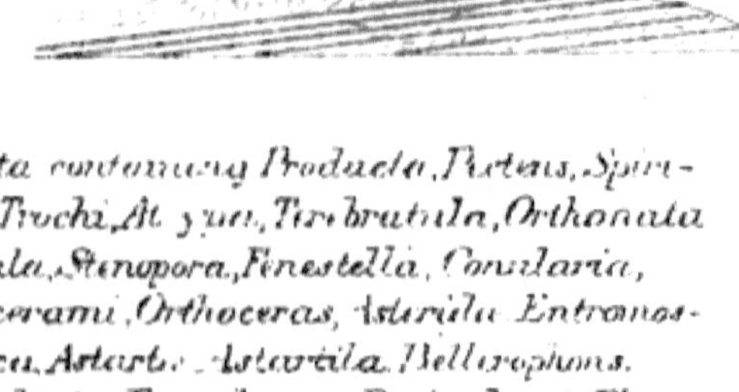

Strata containing Producta, Pectens, Spirifers, Trochi, Atrypa, Terebratula, Orthonota costata, Stenopora, Fenestella, Conularia, Inocerami, Orthoceras, Asterias, Entomostraca, Astarte, Avicula, Bellerophons, Cardinia, Eurydesma, Pachydomus, Pleurotomaria, Terebratula, Glossopteris, Phyllotheca, Noeggerathia, &c. &c.

Lower Coal Measu
with Lepidodendr
Knorria, Calamite
Otopteris, Phylloth
Conularia, Fenestell
Producta and Cri
dea, &c., with thick
of Crinoidal Lim
and Sandstone,
Marine Fossil Fo

Nº 6.

OORAL, *in the* COUNTY *of* CLOUCESTER *to illustrate the relative* . *and some of the* **FOSSIL FLORA** *and* **FAUNA** *found in* *herewith.*

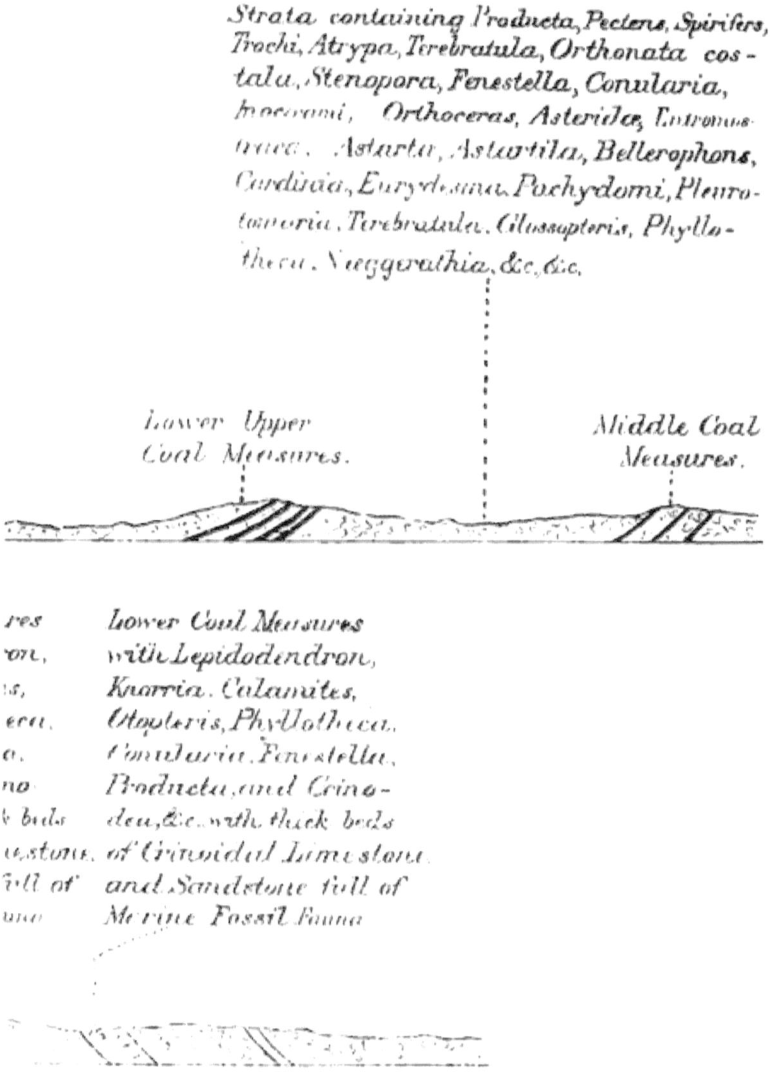

Strata containing Producta, Pectens, Spirifers, Trochi, Atrypa, Terebratula, Orthonata costata, Stenopora, Fenestella, Conularia, Inocerami, Orthoceras, Asteridæ, Entomostraca, Astarta, Astartila, Bellerophons, Cardinia, Eurydesma, Pachydomi, Pleurotomaria, Terebratula, Glossopteris, Phyllotheca, Nœggerathia, &c. &c.

Lower Upper Coal Measures.

Middle Coal Measures.

res	*Lower Coal Measures*
ron,	*with Lepidodendron,*
s,	*Knorria. Calamites,*
eca.	*Otopteris, Phyllotheca.*
a.	*Conularia. Fenestella.*
no	*Producta, and Crino-*
k beds	*dea, &c. with thick beds*
estone.	*of Crinoidal Limestone*
ill of	*and Sandstone full of*
una	*Marine Fossil Fauna*

and rapidly increased; and when the proposed extra shipping appliances are completed at Newcastle and vessels can have rapid despatch, our trade will undoubtedly increase at a much greater rate than it is even now doing.

Sections taken across the Lower Coal Measures at Stroud, in the County of Gloucester.

I annex longitudinal sections (see sections Nos. 17, 18, 19) of the Ward River, Karuah River, and Mamme Johnson strata and seams of coal, and a longitudinal sections (section No. 20) of the provings made at Smith's Creek by the Australian Agricultural Company, in which we find Lepidodendron, Knorria, Calamites, Stigmaria, Otopteris, Phyllotheca, Conularia, Fenestella, Producta, and Crinoidea, &c, with thick beds of crinoidal limestone and sandstone full of palæozoic fossil fauna resting on Silurian rocks.

A on Plan No 5

NORTHERN DISTRICT.

Section showing the Strata and Seams of Coal at A on Plan, at Newcastle, County of Northumberland, New South Wales.

Feet	In.	Description
		No. 1.
41	[illegible]	Soil and Clay
3	[illegible]	Coal
2	[illegible]	Fireclay
6	0	Coal — *Worked by the Borehole Company*
[illegible]	[illegible]	Dark Fireclay with Ironstone Balls
[illegible]	1	Sandstone and Shale, with a few Ironstone Bands
1	9	Very hard Conglomerate
6	5	Blue Shale
1	7	Hard Conglomerate
72	0	Hard Sandstone, with Shale partings, generally a brown Conglomerate
		No. 2.
[illegible]	[illegible]	White metal
[illegible]	[illegible]	Coal, with numerous Shale partings, and Glossopteris, Phyllotheca, &c.
[illegible]	[illegible]	Coal
[illegible]	[illegible]	Fireclay
[illegible]	[illegible]	Coal — *Worked by the Borehole Company*
[illegible]	[illegible]	To 14 feet of white Metal, and Clay, with Glossopteris, Phyllotheca, &c.
[illegible]	0	To 75 feet of Indurated Clay Stone, with Shale partings, and Glossopteris, Phyllotheca, &c.
		No. 3.
[illegible]	[illegible]	Coal
[illegible]	[illegible]	Grey Metal
[illegible]	[illegible]	Coal
[illegible]	[illegible]	Metal
[illegible]	[illegible]	Coal and Shale
[illegible]	[illegible]	Blue Shale
[illegible]	[illegible]	Coal and Shale

Section—*continued.*

Feet	In.	Description
[illegible]	0	Clay Rock (sometimes Grey Post) Metal and Shale, with Glossopteris, Phyllotheca, trunks, &c., of Conifers, and seed vessels of Conifers
		No. 4.
[illegible]	[illegible]	Shale, with 2 inches of Coal
[illegible]	[illegible]	Coal
[illegible]	[illegible]	Fireclay
[illegible]	[illegible]	Coal
[illegible]	[illegible]	Shale
[illegible]	[illegible]	Coal
[illegible]	[illegible]	Coal and Shale
[illegible]	[illegible]	Dark Shale, with partings of Coal
		Dirty Seam. This divides in some places at Newcastle, and forms two coal seams a considerable distance apart.
32	8	Shale, with Glossopteris, Phyllotheca, &c.
		No. 5.
1	[illegible]	Coal
0	1	Band
1	1	Coal
		Yard Seam.
77	10	Rock and Shale
3	11	Coal
[illegible]	[illegible]	Rock and Shale, with Glossopteris, Phyllotheca, &c.

Section—*continued.*

Feet	In.	Description
		No. 6.
1	4	Coal
[illegible]	[illegible]	Band
[illegible]	5	Coal
[illegible]	10	Inferior Coal, with Bands
1	2	Coal
[illegible]	10½	Inferior Coal, with Bands
3	1	Coal

Worked by Australian Agricultural, Co-operative, Lambton, New Lambton, Minmi, Wallsend, and Waratah Companies.

The measurements, from No. 4 to No. 6 inclusive are taken from the A. A. Company's [illegible] Pit.

Below this, a Series of Seams containing several seams of Coal, with Phyllotheca, Vertebraria, Conifers, Glossopteris, &c. &c.; and immediately below the lowest seam of Glossopteris, Sandstones, Grits, and Shales, containing the Fossil Fauna of the Lower Carboniferous formation, and excellent seams of Coal and Cannel Coal.

Section No. 8

B on Plan No. 5

NORTHERN DISTRICT.

SECTION showing the Strata and Seams of Coal at the Wallsend Colliery, near Newcastle, County of Northumberland, New South Wales.

Section—*continued.*

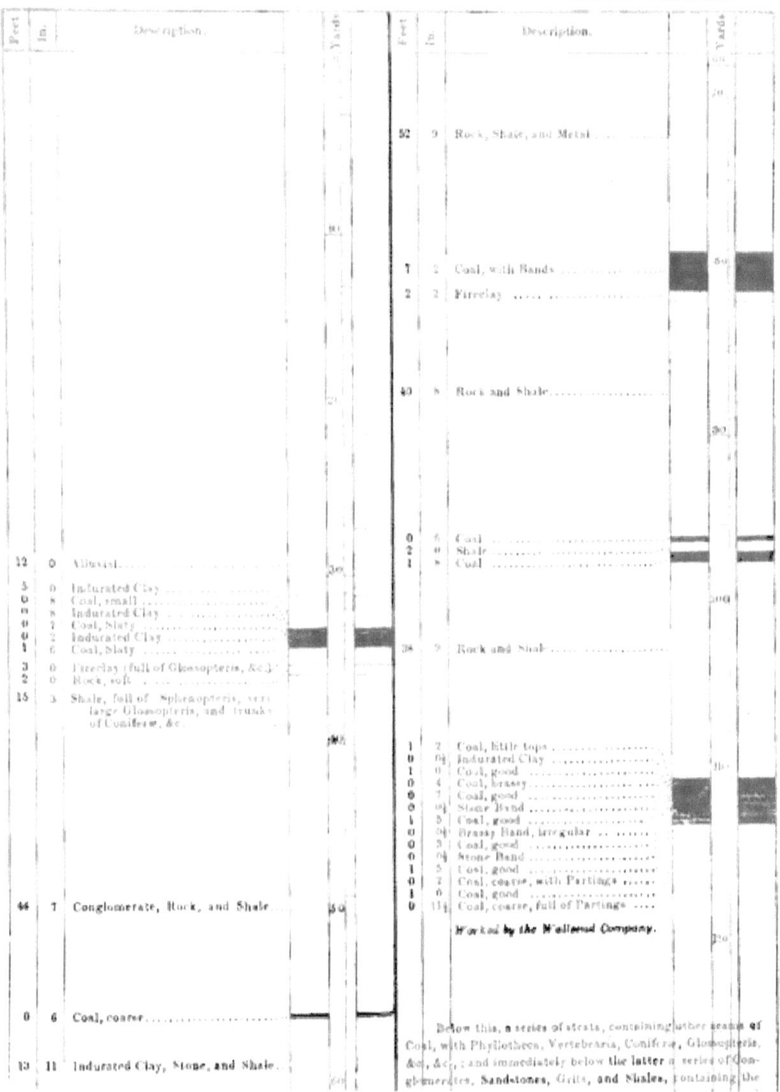

Feet	In.	Description.
12	0	Alluvial
5	0	Indurated Clay
0	8	Coal, small
0	8	Indurated Clay
0	7	Coal, Slaty
0	2	Indurated Clay
1	6	Coal, Slaty
3	0	Fireclay (full of Glossopteris, &c.)
2	0	Rock, soft
15	3	Shale, full of Sphenopteris, very large Glossopteris, and trunks of Coniferæ, &c.
44	7	Conglomerate, Rock, and Shale
0	6	Coal, coarse
13	11	Indurated Clay, Stone, and Shale

Feet	In.	Description.
52	9	Rock, Shale, and Metal
7	2	Coal, with Bands
2	2	Fireclay
40	8	Rock and Shale
0	6	Coal
2	0	Shale
1	8	Coal
38	9	Rock and Shale
1	2	Coal, little tops
0	0½	Indurated Clay
1	0	Coal, good
0	4	Coal, brassy
0	7	Coal, good
0	0½	Stone Band
1	5	Coal, good
0	0½	Brassy Band, irregular
0	3	Coal, good
0	0½	Stone Band
1	5	Coal, good
0	2	Coal, coarse, with Partings
1	6	Coal, good
0	11½	Coal, coarse, full of Partings

Worked by the Wallsend Company.

Below this, a series of strata, containing other seams of Coal, with Phyllotheca, Vertebraria, Coniferæ, Glossopteris, &c., &c.; and immediately below the latter a series of Conglomerates, Sandstones, Grits, and Shales, containing the

WESTERN DISTRICT.

SECTION showing the Strata and Seams of Coal at the Wolgan, County of Cook, New South Wales.

Feet	In.	Description
10	0	**Alluvial**
20	0	**Rock**
3	6	Coal
1	9	Black Metal
		About 3,000 feet over sea level.
13	2	Indurated Claystone with Vertebraria
0	9	Blue Metal and Coal
11	6	Grey Conglomerate, with Shale Bands
1	6	Black Metal, full of stems and plant impressions
0	7	Indurated Clay
2	2	Black Metal and Coal
0	5	Indurated Clay
1	4	Black Metal
0	1	Clay Band
1	0	Black Metal
0	2	Clay Band
0	5	Black Metal
1	0	Fireclay
2	8	Black Metal, with thin layers of Coal
0	7	Fireclay, with Vertebraria
0	1½	Black Metal
0	4	Coal
0	4	Fireclay
0	5	Black Metal
0	8	Coal
0	3	Black Metal
0	10	Coal
1	7	Black Metal, with very few layers of Coal
1	2	Fireclay
1	8	Shale and Indurated Clay
0	8	Black Metal
0	7	Coal
0	6	Black Metal
0	3½	Coal
0	1	Fireclay
2	9	Coal
0	2½	Black Metal
0	8	Fireclay
0	5½	Black Metal
0	9	Coal and Black Metal
0	3	Clay Band
0	1½	Black Metal
1	2	Coal and Indurated Clay
0	3	Black Metal
0	5	Coal and Clay
0	3	Coal (Bituminous)
0	2	Indurated Clay
3	8	Coal (appears to be Bituminous)
1	7	Fireclay and Shale
0	4	Black Metal
0	7	Coal
0	5½	Indurated Clay
0	8	Coal
0	8	Black Metal
0	10	Indurated Clay
9	6	Blue Shale, with Phyllotheca, &c.
15	0	Blue Shale, with Phyllotheca, Vertebraria, Glossopteris, and plant impressions
0	4	Coal

Section—*continued.*

Feet	In.	Description
7	5	Blue Metal and Shale, with Phyllotheca, &c.
0	2½	Indurated Clay
0	8	Coal
0	5	Indurated Clay
0	3	Coal
0	4	Black Metal
0	3	Coal
0	9	Black Metal
0	3	Indurated Clay
0	4	Coal
0	1½	Black Metal
0	1	Indurated Clay
0	6	Coal
0	1	Indurated Clay
0	2½	Black Metal
0	2½	Indurated Clay
0	5	Coal
0	6	Black Metal
0	3½	Coal
0	1	Indurated Clay
0	8	Black Metal and Coal
0	5	Indurated Clay
0	3	Stone
2	0	Fireclay
1	5	Black Metal
1	6	Indurated Clay
1	0	Coal
7	3	Black Metal, Shale, and Indurated Clay
34	0	Brown Sandstone and Conglomerate
25	0	Blue Indurated Claystone and Blue Shale, full of plant impressions, Glossopteris, and Vertebraria
9	6	Very Friable Shale and Indurated Clay
0	3	Coal
0	2	Indurated Clay
0	5	Coal
0	5	Indurated Clay
0	5	Coal
5	9	Blue Shale, full of Vertebraria and Glossopteris
3	3	Indurated Clay and Shale
0	6	Coal
3	6	Shale, Sandstone, and Bands of Indurated Clay
15	6	Soft Brown Sandstone
50	0	Shale and Bands of Clay
85	0	**Various Strata (covered with alluvial), outcrop of a seam of Coal showing in side of Road**

Section—*continued.*

Feet	In.	Description
60	0	**Various Strata (covered with alluvial), outcrop of a seam of Coal not proved**
[illegible]	0	Various Strata (covered with alluvial)
2	0	Coal and Shale
3	0	Friable Shale
[illegible]	3	Coal
5	4	Friable Shale, with Sandstone and Ironstone Bands
6	8	Coal, appears to be about
[illegible]	0	Blue Shale
[illegible]	8	Splint and Bituminous Coal
[illegible]	0½	Clay Band
[illegible]	0	Splint and Bituminous Coal
1	4	Fireclay
[illegible]	0	Sandstone and Conglomerate with Shale

Immediately below this a series of strata of Sandstone, Shale, drifts, and Conglomerate, which contain the Fossil Fauna of the lower Carboniferous formation.

Section 10.

D on Plan No. 5.

WESTERN DISTRICT.

Section showing the Strata and Seams of Coal at Lithgow Valley, County of Cook, New South Wales, taken on land of Thos. Brown, Esq., M.L.A.

Feet	In.	Description
[illegible]	[illegible]	Alluvial
	[illegible]	[illegible]
[illegible]	[illegible]	Coal [illegible]
		[illegible]
[illegible]	[illegible]	Sandstone
[illegible]	[illegible]	Coal with [illegible]
[illegible]	[illegible]	Shale, Sandstones, &c. [illegible]

Section—*continued*.

Feet	In.	Description
[illegible]	[illegible]	Coal [illegible]
[illegible]	[illegible]	Shale and Fireclay
[illegible]	[illegible]	Fireclay
[illegible]	[illegible]	Coal
[illegible]	[illegible]	Friable Shale with Ironstone Bands
[illegible]	[illegible]	Sandstone and Shale
[illegible]	[illegible]	Coal
[illegible]	[illegible]	[illegible] Clay [illegible]
[illegible]	[illegible]	Conglomerate
[illegible]	[illegible]	Coal
[illegible]	[illegible]	Fireclay [illegible]
[illegible]	[illegible]	Grey Sandstone and Conglomerate [illegible]
[illegible]	[illegible]	[illegible] and Shale
[illegible]	[illegible]	Coal
[illegible]	[illegible]	Band
[illegible]	[illegible]	Coal
[illegible]	[illegible]	Band
[illegible]	[illegible]	Coal
[illegible]	[illegible]	Band
[illegible]	[illegible]	Coal
[illegible]	[illegible]	Band
[illegible]	[illegible]	Coal
[illegible]	[illegible]	Band [illegible]
[illegible]	[illegible]	Coal
[illegible]	[illegible]	Band
[illegible]	[illegible]	Coal

Worked by the [illegible]

[illegible] distance below this a series of strata [illegible] Sandstones, Shales, and Grits, containing the [illegible] of the Lower [illegible]

E ON PLAN No. 5.

WESTERN DISTRICT.

SECTION showing the Strata and Seams of Coal at Burragorang, near the junction of the Burragorang and Cox's Rivers, in the County of Cook. New South Wales.

Feet	In.	Description.
12	0	Alluvial
18	0	Rock
4	0	Coal, not proved over 3 feet. *About 250 or 300 feet over Sea level.*
53	0	Conglomerate
0	8	Coal
6	0	Metal and Shale
2	0	Splint Coal
12	9	Indurated Clay, Stone, &c.
2	4	Coal
28	4	Shale with a little Sandstone
2	8	Coal
1	10	Indurated Clay
4	8	Coal and Bands
3	0	Indurated Clay
0	4	Flintstone
1	6	Coal
0	2	Flintstone
2	7	Coal
1	10	Black Stone
2	4	Coal
0	2	Band
7	6	Coal
0	5	Black Stone
0	4	Coal
2	0	Fireclay, full of plant impressions
17	9	Conglomerate and Shale
0	8	Bituminous Shale
1	0	Fireclay

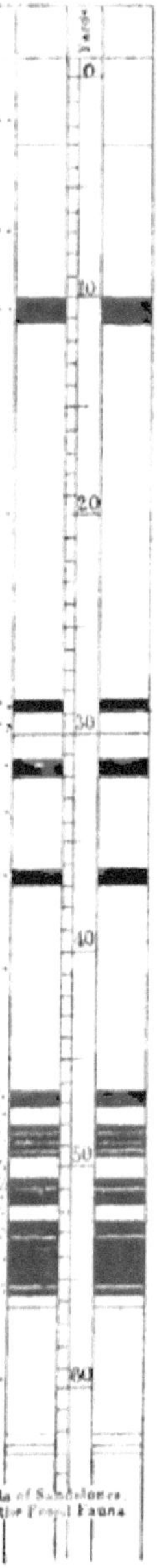

About 200 feet below this a series of strata of Sandstones, Shales, Grits, and Conglomerates, containing the Fossil Fauna of the Lower Carboniferous formation.

Section No. 12

F on Plan No. 5.

SOUTHERN DISTRICT.

Section showing the Strata and Thick Seam of Coal on the Nattai River, near the Fitzroy Iron Mines, in the County of Camden, New South Wales.

Feet	In.	Description.
		Sandstone ....
2	7	Coal (fine grained splint) ....
0	3	Black bass ....
0	3	Coal ....
1	4	Sandstone ....
5	6	Splint Coal ....
0	2	Black Metal........................
5	0	Grey Rock ....
1	1	Splint Coal ....
0	4½	Coal and dirt ....
2	2	Splint Coal ....
0	2	Coal and dirt ....
4	6	Splint Coal ....
0	0½	Parting ....
7	6	Splint and bituminous looking Coal ....
0	3	Stone, &c. ....
1	3	Coal ....
0	2	Coal and dirt ....
3	6	Coal ....
0	0½	Parting ....
0	2	Coal ....
0	2	Parting ....
0	5	Coal ....
0	1	Parting ....
0	7	Coal ....
0	2	Parting ....
0	11	Coal ....
0	2	Stone ....
3	9	Grey Rock ....
15	0	Sandstone ....
5	0	Fine Conglomerate ....
0	9	Grey Conglomerate ....
11	0	Sandstone ....
15	0	Metal ....
13	0	Conglomerate........................
18	0	Indurated Claystone ....
[illegible]	0	Greenstone ....

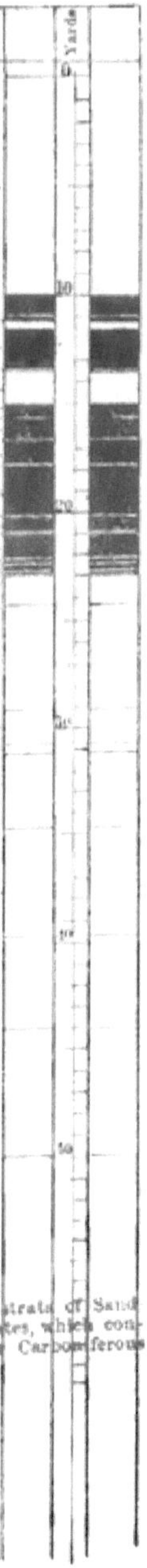

Immediately below this a series of strata of Sandstones, Shales, Grits, and Conglomerates, which contain the fossil Fauna of the lower Carboniferous formation.

Section N°13

G on Plan N°5

SOUTHERN DISTRICT.

Section showing the Strata and Seams of Coal at Kangaroo Creek, near in the County of Camden, New South Wales.

Feet.	In.	Description.
10	0	Alluvial ...
20	0	Rock ...
3	0	Shale ...
4	6	Coal—5 feet to ...
3	0	Rock ...
4	0	Coal and Shale ...
12	0	Coal with bands ...
2	0	Black Metal, full of Vertebraria and Coal ...
5	0	Grey Metal and Fireclay with Glossopteris ...
		Conglomerate ...

A short distance below this a series of strata of Conglomerates, Sandstones, Shales, and Grits, containing the [illegible] fossils of the lower Carboniferous formation [illegible] underlying the latter is Granite.

Section No. 14.

H on Plan, No. 5.

SOUTHERN DISTRICT.

SECTION showing the Strata and Seams of Coal at Mt. Keira, near Wollongong, County of Camden, New South Wales.

Feet.	In.	Description.
15	0	Alluvial
10	0	Rock
7	6	Coal
6	0	Parting
		This seam of Coal is 80 feet above sea level.
31	0	Brown and Grey Sandstone
4	0	Coal
59	0	Blue Metal and Grey Post
9	0	Bituminous Shale or Inferior Coal
0	5	Parting
1	0	Bituminous Shale or Inferior Coal
0	5	Parting
0	11	Bituminous Shale or Inferior Coal
0	1	Parting
1	[illegible]	Select Coal
[illegible]	0	Shale
[illegible]	[illegible]	Brown Sandstone
[illegible]	[illegible]	Black Metal
[illegible]	10	Coal (bright, [illegible])
[illegible]	[illegible]	Black and Blue Metal
[illegible]	[illegible]	Coal
[illegible]	[illegible]	Parting
[illegible]	[illegible]	Coal
[illegible]	[illegible]	Parting
[illegible]	[illegible]	Coal
[illegible]	[illegible]	Blue Metal
[illegible]	[illegible]	Coal
[illegible]	[illegible]	Grey Metal
[illegible]	[illegible]	Coal
[illegible]	[illegible]	Parting
[illegible]	[illegible]	Coal
[illegible]	[illegible]	Black Stone
[illegible]	[illegible]	Coal
[illegible]	[illegible]	Freestone (full of Vertebraria, &c.)
12	[illegible]	Grey Sandstone
10	[illegible]	Brown and Grey Sandstone
4	0	Coal, Shale, and Stone
15	0	Blue Metal with Glossopteris
1	3	Brown Sandstone
0	[illegible]	Coal
0	1	Parting
1	6	Coal
0	8	Dark Metal (full of Glossopteris)
0	10	Coal
11	0	Grey Shale with Ironstone Bands (Full of Glossopteris and vegetable Impressions)

Section—*continued.*

Feet.	In.	Description.
[illegible]	7	Brown Sandstone
[illegible]	[illegible]	Friable Grey Shale
16	0	Fine Grey Sandstone in thin layers, full of Glossopteris and vegetable impressions
[illegible]	[illegible]	Ironstone
[illegible]	[illegible]	Hard stone with [illegible] and impressions of plants
[illegible]	[illegible]	Ironstone, very [illegible]
[illegible]	[illegible]	Metal, with Glossopteris in top part
[illegible]	[illegible]	Coal, &c., not properly proved
[illegible]	[illegible]	Band
[illegible]	[illegible]	Coal, &c., not properly proved
[illegible]	0	Shale and Metal
[illegible]	0	Coal
[illegible]	0	Blue Metal
79	0	Grey Post
80	0	Hard Rock
[illegible]	0	Brown Sandstone

Section—*continued.*

Feet.	In.	Description.
90	0	Grey Post
[illegible]	[illegible]	Indurated Clay
[illegible]	[illegible]	Coal and Shale
[illegible]	[illegible]	Indurated Clay
[illegible]	[illegible]	Coal
[illegible]	[illegible]	Parting
[illegible]	[illegible]	Coal
[illegible]	[illegible]	Shale
[illegible]	[illegible]	Coal
[illegible]	[illegible]	Shale full of Glossopteris
[illegible]	[illegible]	Coal
[illegible]	[illegible]	Shale full of Glossopteris
[illegible]	[illegible]	Coal
[illegible]	[illegible]	Metal with Glossopteris
[illegible]	[illegible]	Metal with Glossopteris
14	[illegible]	Sandstone
[illegible]	[illegible]	Coal
[illegible]	[illegible]	Coal and Shale
[illegible]	0	Sandstone, &c., with trunks of Conifers lying nearly horizontal
[illegible]	[illegible]	Bituminous Shale
[illegible]	[illegible]	Shale
[illegible]	0	Conglomerate and Sandstone
[illegible]	1	Metal mixed with Ironstone full of Glossopteris, with trunks of Conifers lying horizontal
[illegible]	[illegible]	Grey and Black Shale with Metal in nodular masses and layers full of Vertebraria and Glossopteris
[illegible]	[illegible]	Shale and Sandstone
[illegible]	0	Sandstone
[illegible]	0	Shale and Metal

Below this a series of Strata of Sandstone, Shales, Grits, and Conglomerates, the upper portion of which have contain Fossil Flora and Fauna, and about [illegible] feet lower are the beds containing the Marine Fossil Fauna of the Lower Carboniferous formation.

SOUTHERN DISTRICT.

SECTION showing the Strata and Seams of Coal formerly belonging to A. T. Holroyd, Esq., and others, about 4 miles south-west of Coal Cliff, in the parish of Southend, County of Cumberland, New South Wales.

Feet.	In.	Description.
13	0	Alluvial
15	0	Rock
40	0	Sandstone
[illegible]	[illegible]	Ironstone, &c.
[illegible]	[illegible]	Sandstone
[illegible]	[illegible]	Ironstone, &c.
[illegible]	[illegible]	Shale
5	0	Coal
3	0	Shale, with Vertebraria
13	0	Sandstone and Conglomerate
2	0	Coal (fine grained splint) Not properly opened out to ascertain whether this is the full thickness
20	0	Shale and Sandstone, with Iron-stone Bands
5	0	Coal, Fireclay, and Shale
6	0	Coal, with Partings
1	0	Shale and Ironstone
6	0	Coal, with numerous Bands
[illegible]	[illegible]	Coal
[illegible]	[illegible]	Rock
[illegible]	[illegible]	Fireclay
[illegible]	[illegible]	Coal
[illegible]	[illegible]	Shale and Coal
[illegible]	[illegible]	Bituminous Shale and Rock nodular Ironstone
[illegible]	[illegible]	Coal and Shale
[illegible]	[illegible]	Fireclay
[illegible]	[illegible]	Coal
[illegible]	[illegible]	Bituminous Shale and Fireclay
[illegible]	[illegible]	Dark Metal and Shale, full of Vertebraria
8	0	Grey Sandstone
0	1	Impure Limestone
10	0	Grey Post with hard veins

Section—continued.

Feet.	In.	Description.
[illegible]	[illegible]	Coal
[illegible]	[illegible]	Parting
[illegible]	[illegible]	Coal, with Bituminous Partings
[illegible]	[illegible]	Coal and Partings
[illegible]	[illegible]	Coal
[illegible]	[illegible]	Dark Metal
[illegible]	[illegible]	Laminated Fireclay
[illegible]	[illegible]	Dark Bituminous Shale, with Vertebraria
[illegible]	[illegible]	Shale, with Bands of Ironstone
[illegible]	[illegible]	Sandstone, with Bands of Ironstone
[illegible]	[illegible]	Grey Sandstone, with thin streaks of vegetable matter
[illegible]	[illegible]	Coal with Partings (about)
[illegible]	[illegible]	Ironstone
[illegible]	[illegible]	Bright Bituminous Coal
		Sandstone

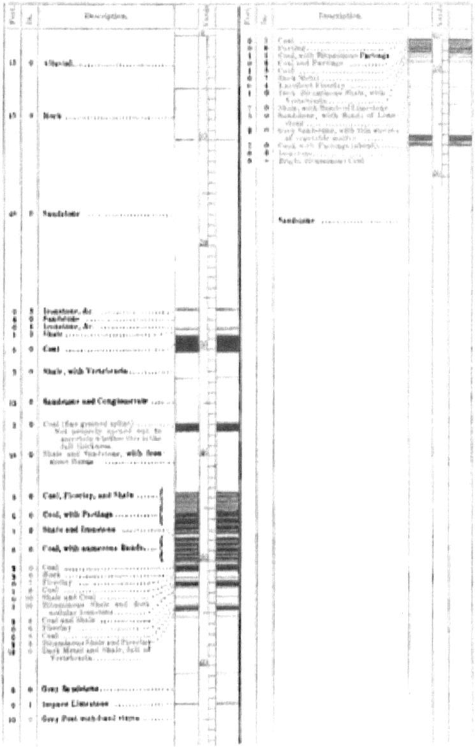

Ion Plan No. 5

SOUTHERN DISTRICT.

Section showing the Strata and Seams of Coal at Coal Cliff, Parish of Southend, County of Camden, belonging to His Honor Mr. Justice Hargraves, and adjoining the extensive Mineral Leases of Alexander Stuart, Esq.

Feet	In.	Description
[illegible]	0	Alluvial
15	0	Rock
5	5	[illegible]
10	[illegible]	[illegible]
6	0	Coal, fine-grained Splint Coal with Partings of [illegible]
8	0	[illegible] feet [illegible] inches of Fireclay
20	0	Sandstone and Shale
2	4	Coal, very fine grained Splint Coal
[illegible]	[illegible]	Fireclay
28	6	Sandstone and Shale
3	0	Coal
9	8	Sandstone and Shale
1	8	Coal
21	6	Sandstone
5	6	Shale
[illegible]	6	Coal
[illegible]	6	Shale
[illegible]	3	Coal
[illegible]	1	Band
[illegible]	[illegible]	Coal
[illegible]	6	Rock and Shale
[illegible]	[illegible]	Coal
[illegible]	[illegible]	Band
[illegible]	[illegible]	Coal
[illegible]	2	Band
[illegible]	5	Coal
[illegible]	2	Band
[illegible]	[illegible]	Coal
[illegible]	[illegible]	Band
[illegible]	[illegible]	Coal
[illegible]	1	Band
[illegible]	3	Coal
[illegible]	[illegible]	Metal and Fireclay
[illegible]	0	Various Strata
0	8	Coal
33	0	Various Strata

Section—*continued*

Feet	In.	Description
2	8	Coal with a parting

[illegible]

Section—*continued*

Feet	In.	Description
[illegible]	[illegible]	Coal
[illegible]	[illegible]	Band
[illegible]	[illegible]	Coal
[illegible]	[illegible]	Band
[illegible]	[illegible]	Coal
[illegible]	[illegible]	Band
[illegible]	[illegible]	Coal
[illegible]	[illegible]	Band
[illegible]	[illegible]	Coal
[illegible]	[illegible]	Band
[illegible]	[illegible]	Coal
[illegible]	[illegible]	Band
[illegible]	[illegible]	Coal

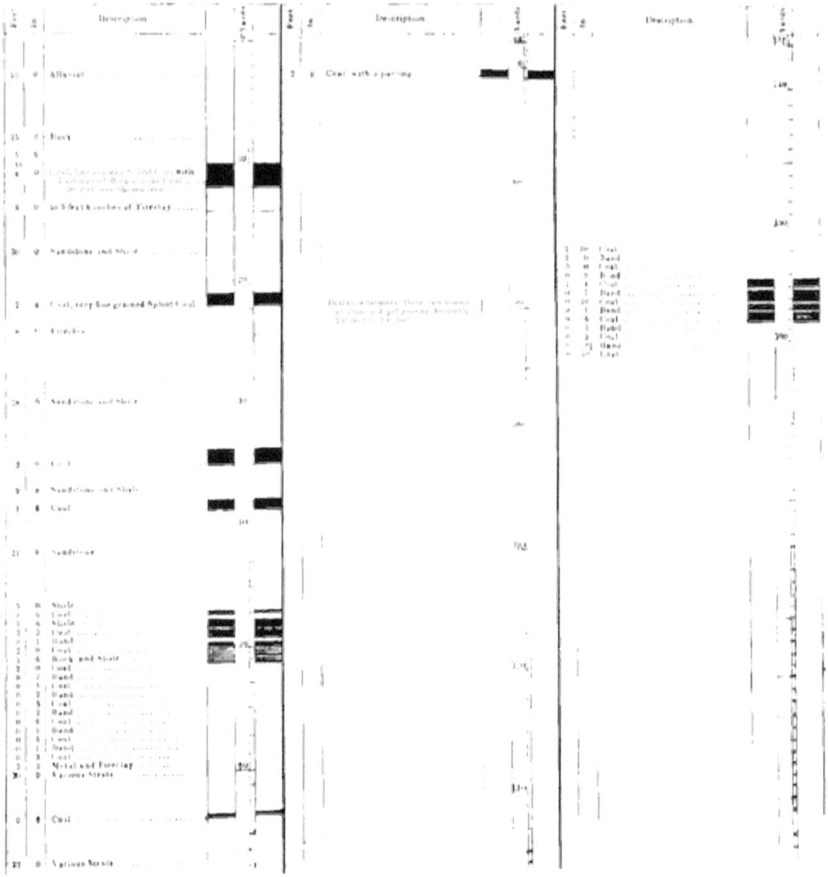

SECTION OF LOWER COAL MEASURES AT WARD RIVER, ON THE AUSTRALIAN AGRICULTURAL COMPANY'S PROPERTY, IN THE COUNTY OF GLOUCESTER, NEW SOUTH WALES.

SECTION Nº [illegible]

Bare Shale

SECTION OF THE LOWER COAL MEASURES AT [illegible] KARUAH RIVER, ON THE AUSTRALIAN AGRICULTURAL COMPANY'S **PROPERTY NEAR STROUD**, IN THE COUNTY OF [illegible]

SECTION Nº [illegible]

SECTION Nº [illegible]

LONGITUDINAL SECTION TAKEN AT RIGHT ANGLES TO THE STRIKE NEARLY EAST WEST AT [illegible]

MAMMIE JOHNSON

K L M N O

SECTION Nº 20

Section 8000 feet in length taken along Frontage made by the Australian Agricultural Company across a portion of the lower Coal Measures at Smiths Creek, near Stroud, in the County of Gloucester.

APPENDIX.

DESCRIPTION of Coal Seams at Mount Kembla, near Wollongong, by R. W. Moody, Esq., Mining Engineer.

LOCALITY, &c.

This coal field is situated about 5 miles in a south-westerly direction from Wollongong, in the county of Cumberland, New South Wales, being on the same line of ranges as the Bulli, Mount Pleasant, and Mount Keira Coal properties, and the top seam in Mount Kembla series, being identical with the celebrated steam coal seam worked so successfully at those collieries for so many years.

The various coal seams, as per "section of measures" appended, are seen cropping out in the slopes of the Mount Kembla Ranges, and are found very nearly level, rising slightly into the mountain, and are at very convenient distances from each other, the bottom seam being nearly on a level with the lands, over which a line of permanent railway may be very economically constructed, by means of which the produce of the mines, &c. may be conveyed to the port of Wollongong, for shipment at the Government coal shoots already in use there by the Mount Pleasant and Mount Keira Coal Companies; or a private shipping place could easily be constructed about a mile south of Wollongong, at a moderate cost, which I believe would meet all the requirements of a very large output, besides being able to form a junction with the proposed "Sydney and Illawarra" Line of Railway, which I have little doubt will soon be commenced, and which will admit of the produce of the mines being sent to Sydney for shipment and for home consumption.

QUANTITIES—NO. 1 SEAM.

The highest in the series, or "Bulli Seam," is 7 feet thick of fine clean coal without any bands or refuse, in fact I have not seen this seam so good in any part of the Wollongong District, the quality of which, especially for steam purposes, is I think too well known to need any comment in a report of this kind.

I believe that at least 500 acres of the estate will contain this fine seam of coal, which I estimate will yield 5,285,000 tons of marketable coal—sufficient for an output of 500 tons per day for over thirty-five years.

No. 2 Seam.

Lying about 35 feet below No. 1, is 4 feet thick, is a fine bituminous or house-fire coal, and which I calculate will extend underneath at least 520 acres, and will therefore yield about 3,140,000 tons of coal, sufficient for an output of 500 tons per day for over twenty-one years.

No. 3 Seam.

Lying about 125 feet below the last or No. 2 seam; this seam is 17 feet thick in all, although 7 feet only has yet been opened out. The bottom portion only which has been proved, is a very fine semi-bituminous free-burning coal, giving out a great heat, leaving little ash and no clinkers, and which I believe will be found to answer well for house, steam, and gas purposes; and although only 7 feet of it has been tried, I am disposed to believe from appearances at least 14 feet will be found workable and of a good marketable quality, thus leaving 3 feet for bands or refuse, and I believe this fine seam will underlie some 550 acres, and will therefore yield over 11,627,000 tons of coal, sufficient for an output of 500 tons per day for nearly seventy-eight years.

No. 4 Seam.

Lying about 45 feet below No. 3 Seam, and is an exceedingly rich and valuable kerosene oil shale seam of 4 feet 9 inches thick, 2 feet 3 inches on the upper portion being of a very superior quality, yielding 50 gallons of oil to the ton. The bottom portion, 2 feet 6 inches thick, is not so rich in quality, and has not yet been fairly tested, yet it has been worked with the other to make sufficient height in the mine to ensure economy in working. I may remark that this oil shale has been explored for about 120 yards in from the outcrop, and is gradually increasing in thickness and quality, the kerosene works now in full operation on the estate being supplied from this mine. I estimate that this seam, taking only the 2 feet 3 inches of proved quality will yield in the aggregate something like 1,450,000 tons of shale, producing 72,500,000 gallons of refined oil, or about 100,000 gallons per year for over seventy-two years.

No. 5 Seam.

Lying about 40 feet below the kerosene shale, is 7 feet in thickness, and has also been sufficiently explored to prove its quality. This seam has a band of light blue shale running through the middle of it, leaving 6½ feet of good marketable coal, the upper portion of which is a very fine semi-bituminous coal, and the lower portion below the band is a very rich close-grained steam coal, forming in the whole a very valuable and useful seam of coal 6½ feet thick of clean marketable coal. I consider this seam will underlie 590 acres of the estate, and will therefore yield 5,315,400 tons of good coal, which will supply a vend of 500 tons a day for over thirty-six years.

No. 6 Seam.

Of the series lying about 40 feet below the No. 5 Seam, consists of two beds of coal, together 14 feet thick; the lower portion or 7 feet only have as yet been opened out; this portion has three thin bands of shale running through it, together about 5 inches, reducing the thickness of marketable coal to 6½ feet. This is a fine semi-anthracite coal, with several very thin bands of bituminous coal running through the seam, which will I think tend greatly to cheapen its production. The upper 7 feet portion has not yet been tried, but from its outcrop appearance I have no doubt a large portion, if not the whole, will be found a good marketable coal; but taking that portion only which has already been proved, namely 6½ feet, which will underlie the whole of the estate of 600 acres, will yield 5,853,000 tons, which will supply an output of 500 tons a day for forty years.

Aggregate Yield.

The five seams of coal contained in this 600 acres of land will therefore yield the enormous, and I may almost safely state the unprecedented quantity of 31,250,000 tons of coal, which will supply a vend of 1,000 tons a day for over 100 years; this is independent of the exceedingly rich bed of kerosene oil shale, which is sufficient to yield 2,000 gallons of refined oil per week for over seventy-two years.

Winning, &c.—Working.

The position of all the seams that I have mentioned are so favourably situated that the coal from each can be got by tunnelling into the side of the mountain range and conveyed to the proposed railway terminus below by self-acting incline planes.

I append a section of measures taken by barometrical observations.

SECTION OF MEASURES.

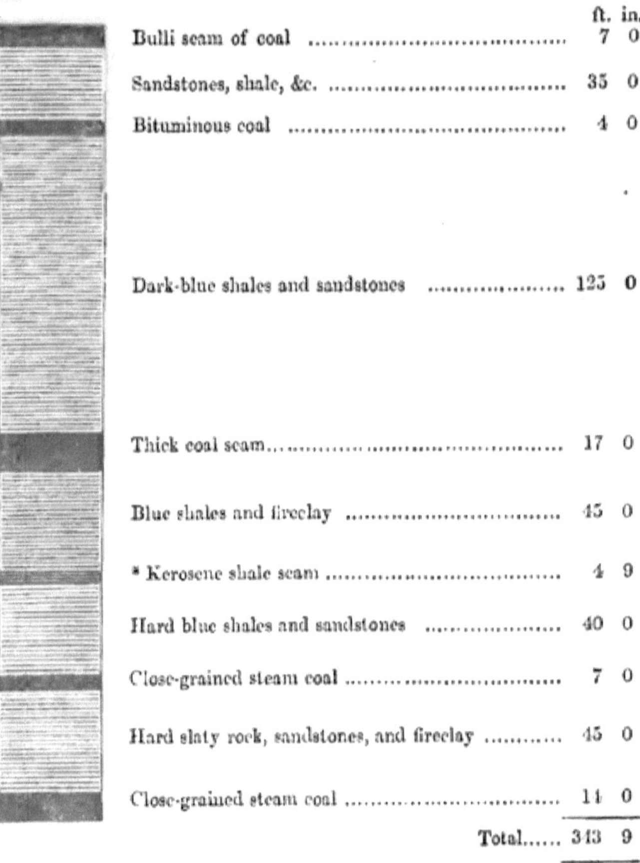

	ft.	in.
Bulli seam of coal	7	0
Sandstones, shale, &c.	35	0
Bituminous coal	4	0
Dark-blue shales and sandstones	125	0
Thick coal seam	17	0
Blue shales and fireclay	45	0
* Kerosene shale seam	4	9
Hard blue shales and sandstones	40	0
Close-grained steam coal	7	0
Hard slaty rock, sandstones, and fireclay	45	0
Close-grained steam coal	14	0
Total	343	9

* 2 ft. 3 in. now being wrought.